Reading and Writing Strategies *for the* Secondary MATHEMATICS Classroom *in a* PLC at Work®

Daniel M. Argentar
Katherine A. N. Gillies
Maureen M. Rubenstein
Brian R. Wise

EDITED BY
Mark Onuscheck
Jeanne Spiller

555 North Morton Street
Bloomington, IN 47404
800.733.6786 (toll free) / 812.336.7700
FAX: 812.336.7790

email: info@SolutionTree.com
SolutionTree.com

Visit **go.SolutionTree.com/literacy** to download the free reproducibles in this book.

Printed in the United States of America

Library of Congress Cataloging-in-Publication Data

Names: Argentar, Daniel M., 1970- author. | Gillies, Katherine A. N., author. | Rubenstein, Maureen M., author. | Wise, Brian R., author. | Onuscheck, Mark, editor. | Spiller, Jeanne, editor.
Title: Reading and writing strategies for the secondary mathematics classroom in a PLC at work / Daniel M. Argentar, Katherine A.N. Gillies, Maureen M. Rubenstein, and Brian R. Wise (authors) ; Mark Onuscheck and Jeanne Spiller (editors).
Description: Bloomington, IN : Solution Tree Press, [2024] | Includes bibliographical references and index.
Identifiers: LCCN 2024027081 (print) | LCCN 2024027082 (ebook) | ISBN 9781945349751 (paperback) | ISBN 9781949539004 (ebook)
Subjects: LCSH: Mathematics--Study and teaching (Middle school)--United States. | Mathematics--Study and teaching (Secondary)--United States. | Reading (Middle school)--United States. | Reading (Secondary)--United States. | Composition (Language arts)--Study and teaching (Middle school)--United States. | Composition (Language arts)--Study and teaching (Secondary)--United States. | Professional learning communities--United States. | Interdisciplinary approach in education--United States.
Classification: LCC QA13 .A74 2024 (print) | LCC QA13 (ebook) | DDC 510.71/273--dc23/eng/20240628
LC record available at https://lccn.loc.gov/2024027081
LC ebook record available at https://lccn.loc.gov/2024027082

Solution Tree
Jeffrey C. Jones, CEO
Edmund M. Ackerman, President

Solution Tree Press

President and Publisher: Douglas M. Rife
Associate Publishers: Todd Brakke and Kendra Slayton
Editorial Director: Laurel Hecker
Art Director: Rian Anderson
Copy Chief: Jessi Finn
Senior Production Editor: Suzanne Kraszewski
Copy Editor: Evie Madsen
Proofreader: Jessi Finn
Text and Cover Designer: Abigail Bowen
Acquisitions Editor: Hilary Goff
Assistant Acquisitions Editor: Elijah Oates
Content Development Specialist: Amy Rubenstein
Associate Editor: Sarah Ludwig
Editorial Assistant: Anne Marie Watkins

ACKNOWLEDGMENTS

Thank you to our significant others and families for supporting us through the process of writing this book. This book wouldn't be possible if we didn't have their continued support and encouragement. Thank you to Brandi, Ami, Nadav, and Alon Argentar; Nick, Cole, and Chase Gillies; Ryan, Camilla, and Kallie Rubenstein; and Erin, Colin, and Julianne Wise.

We are indebted to all of the mentors, teachers, and students who have shaped our thinking throughout our teaching careers. In particular, we are grateful for the partnership and collaborations we have enjoyed with the Adlai E. Stevenson High School mathematics department and D219 Niles Township High Schools. It is their commitment to literacy within our professional learning community that is the foundation for adapting and creating many of the strategies in this book.

Thanks go out to Adlai E. Stevenson High School math teachers Kate Hoopes, Vicki Kieft, Jennifer Parisi, Julisa Ruiz, Andrea MacLennan, Eric Anderson, Carly Hope, Mary Smaga, and Sakthi Shanmugasundaram. Your passion, expertise, inspiration, and support have been invaluable.

Thank you to Niles North High School's many committed literacy teams and to the math teachers who have worked tirelessly to support their students through challenging texts. It has been a pleasure working through the many challenges that are a part of this committed work.

Thank you to the STEM educators and the leaders in the field who have served as thought partners over the years, namely, Ellen Foley, Ankur Joshi, Justin Kleinheider, and Shane Talbott. Your commitment to colleague collaboration, student growth, and teacher preparation is essential.

We also thank the literacy specialists, coaches, and teachers from Downers Grove North High School and the Chicago Area Literacy Leaders (CALL) group for sharing their expertise and inspiring the important literacy work and collaborations we engage in daily.

Thank you to the administrative leaders at Adlai E. Stevenson High School and Niles North High School for their encouragement and support. By prioritizing literacy in these schools, they have allowed our work to grow and inspire our school communities.

Finally, a very special thanks to Mark Onuscheck, director of curriculum, instruction, and assessment at Adlai E. Stevenson High School, whose compassion, humor, inspiration, guidance, and friendship motivate and inspire us regularly. Without his steady leadership, there would be no book to write.

Solution Tree Press would like to thank the following reviewers:

Lindsey Bingley
Literacy and Numeracy Lead
Foothills Academy
Calgary, Alberta, Canada

Paula Mathews
STEM Instructional Coach
Dripping Springs ISD
Dripping Springs, Texas

Rachel Swearengin
Fifth-Grade Teacher
Manchester Park Elementary School
Olathe, Kansas

Sheryl Walters
Instructional Design Lead
Calgary, Alberta, Canada

Visit **go.SolutionTree.com/literacy** to download the free reproducibles in this book.

TABLE OF CONTENTS

Reproducible pages are in italics.

CHAPTER 6

Strategies for Writing About Mathematics

ABOUT THE SERIES EDITORS

Mark Onuscheck is director of curriculum, instruction, and assessment at Adlai E. Stevenson High School in Lincolnshire, Illinois. He is a former English teacher and director of communication arts. As director of curriculum, instruction, and assessment, Mark works with academic divisions about professional learning, articulation, curricular and instructional revision, evaluation, assessment, social-emotional learning, technologies, and Common Core implementation. He worked as a lecturer and adjunct professor at DePaul University for over twenty years.

Mark was awarded the Quality Matters Star Rating for his work in online teaching. He helps to build curriculum and instructional practices for TimeLine Theatre's arts integration program for Chicago Public Schools. Additionally, he is a National Endowment for the Humanities grant recipient and a member of the Association for Supervision and Curriculum Development, National Council of Teachers of English, International Reading Association, and Learning Forward.

Mark earned a bachelor's degree in English and classical studies from Allegheny College and a master's degree in teaching English from the University of Pittsburgh.

Jeanne Spiller is assistant superintendent for teaching and learning for Kildeer Countryside Community Consolidated School District 96 in Buffalo Grove, Illinois. School District 96 is recognized on All Things PLC (www.AllThingsPLC.info) as one of only a small number of school districts where all schools in the district earn the distinction of a model professional learning community. Jeanne's work focuses on standards-aligned instruction and assessment practices. She supports schools and districts across the United States to gain clarity about and implement the four critical questions of professional learning communities. She is passionate about collaborating with schools to develop systems for teaching and learning that keep the focus on student results and help teachers determine how to approach instruction so that all students learn at high levels.

Jeanne received a 2014 Illinois Those Who Excel Award for significant contributions to the state's public and nonpublic elementary schools in administration. She is a graduate of the 2008 Learning Forward Academy, where she learned how to plan and implement professional learning that improves educator practice and increases student achievement. She has served as a classroom teacher, team leader, middle school administrator, and director of professional learning.

Jeanne earned a master's degree in educational teaching and leadership from Saint Xavier University, a master's degree in educational administration from Loyola University, Chicago, and an educational administrative superintendent endorsement from Northern Illinois University.

To learn more about Jeanne's work, visit www.livingtheplclife.com, and follow @jeeneemarie on X (formerly Twitter).

To book Mark Onuscheck or Jeanne Spiller for professional development, contact pd@SolutionTree.com.

ABOUT THE AUTHORS

Daniel M. Argentar is a literacy coach and communication arts teacher at Adlai E. Stevenson High School in Lincolnshire, Illinois. In his previous work as a sixth-grade teacher, he taught reading, language arts, social studies, and science. Since 2001, he has provided academic literacy support to struggling freshmen and sophomores, in addition to teaching other college prep and accelerated English courses. In his coaching role, he partners with instructors from all divisions to increase disciplinary literacy for students—running book studies, professional development sessions, and one-on-one coaching meetings.

Daniel received a bachelor's degree in speech communications from the University of Illinois at Urbana-Champaign, an English teaching degree and a master's degree in curriculum and instruction from Northeastern Illinois University, and a master's degree in reading from Concordia University in Chicago.

To learn more about Daniel's work, follow @dargentar125 on X (formerly Twitter).

Katherine A. N. Gillies is a reading specialist and English teacher at Niles North High School in Skokie, Illinois, where she previously served as a literacy coach. Katherine serves as the lead architect of schoolwide literacy-improvement work, including building a comprehensive system of intervention and support for struggling readers, as well as crafting research-based curricula to ensure continued literacy growth for all students. Katherine leads several collaborative teams and cross-curricular initiatives aimed at using data to inform instruction, building capacity for disciplinary literacy, and employing responsible assessment practices at the secondary level and specializes in building- and district-level literacy-needs analysis and improvement planning. She has presented on these topics at local and national conferences, including for the National Council of Teachers of English. Additionally, she has worked collaboratively across the Chicagoland area to help other area schools develop their research-driven multitiered systems of support and literacy systems and has consulted with various school districts to develop staff literacy capacity across all content areas.

Katherine earned a bachelor's degree in literature and secondary education from Saint Louis University; a master's degree in literacy, language, and culture with a reading specialist certification from the University of Illinois Chicago; and a master's degree in educational leadership and administration from Concordia University Chicago. She is a doctoral candidate at Vanderbilt University's Peabody College of Education. She is also a certified Project CRISS (Creating Independence through Student-owned Strategies) trainer and Orton Gillingham instructor and is recognized as a Distinguished Educator by Renaissance Learning. She was a part of the AIM Coaching Model Design Team, where the University of Texas at Austin and University of Maryland research teams partnered to develop a middle school instructional coaching model with research-based outcomes for students with disabilities. This project was funded by the U.S. Department of Education Institute of Education Sciences.

To learn more about Katherine's work, follow @katherineangillies on Instagram.

Maureen M. Rubenstein is a literacy coach and special education instructor at Adlai E. Stevenson High School. As a teacher, she works with students on individualized education plans who have been diagnosed with reading, writing, and emotional disabilities. In her coaching role, she partners with instructors from all divisions to work on disciplinary literacy. In addition to coaching individual teachers, she works with the other literacy coaches to coordinate and implement book clubs, professional development sessions, and one-to-one coaching sessions.

Maureen received a bachelor's degree in special education from Illinois State University, a master's degree in language literacy and specialized instruction (reading specialist) degree from DePaul University, and a master's degree in educational leadership from Northern Illinois University. Maureen is also a certified trainer for Project CRISS (Creating Independence through Student-owned Strategies), and she is certified to teach Wilson Reading.

To learn more about Maureen's work, follow @SHS_LiteracyMR on X (formerly Twitter).

Brian R. Wise is a literacy coach and English department chair at Deerfield High School in Deerfield, Illinois. As a department chair, he facilitates professional learning, provides supervision and evaluation, and guides curriculum and instruction. He has taught a wide array of English language arts and literacy-intervention courses throughout his teaching career. As a high school literacy coach, Brian worked with faculty members from all content areas to build teachers' capacity for embedding literacy skills into classroom instruction and assessment.

Brian received his bachelor's degree in English education from Boston University, a master's degree in English from DePaul University, and master's degrees in reading and principal preparation from Concordia University, Chicago.

To learn more about Brian's work, follow @Wise_Literacy on X (formerly Twitter).

To book Daniel M. Argentar, Katherine A. N. Gillies, Maureen M. Rubenstein, or Brian R. Wise for professional development, contact pd@SolutionTree.com.

PREFACE

To begin this book, and to immediately demonstrate the value of professional learning communities (PLCs) in supporting positive, thoughtful collaboration, we want to share a real-life experience we had with a group of fellow teachers in our school in our role as literacy coaches. We believe this experience serves as an example of the familiar struggle occurring in many schools when teachers from various content areas strive to approach literacy instruction.

In the spring of 2016, we were approached by the director of our mathematics department. There was an acknowledged need and desire to develop participation in our literacy team with mathematics teachers, but there was also a struggle to create buy-in. Aside from the obvious literacy skills required to read mathematics word problems, mathematics teachers were skeptical that there was a great need for literacy within their discipline. And so, as literacy coaches, we put on our marketing hats, invited mathematics teachers to a learning lunch, planned to show off some important connections between literacy and mathematics, and invited those interested to join our team. Luckily, a small group of teachers agreed to engage in the work. As we had done with other disciplines, we began our work by listening to our mathematics colleagues describe the various issues they encountered with their students. The list of issues was similar to discussions we had with other disciplinary teams, and we compiled the following list of reasons our mathematics colleagues felt students struggled in their classes.

- Students might struggle with mathematics-specific vocabulary and words with multiple meanings.
- Students might have poor knowledge of reading strategies, poor usage of reading strategies, or both.
- Students might have difficulty remembering formulas and concepts and focusing on important details.
- Students might have difficulty making logical inferences.
- Students might struggle to explain how they approach and solve problems.

- Students might struggle with their core writing skills with a need to improve focus, increase the use of evidence, and provide clearer justification, especially when composing proofs.

In addition, we found that mathematics presented two more challenging aspects for our work. First, when we examined the student makeup in many mathematics classes, we noticed a wide variety of initial student literacy levels and a range of multilingual learners. Whereas many courses throughout the school were limited in age (one single grade level) and ability (regular or accelerated), mathematics classes often included multiple ages and students of all learning styles and abilities. In fact, some mathematics sections may have students with reading abilities ranging from low English proficiency to elementary to post–high school. The team had their work cut out for them in supporting such a wide variety of students.

The other challenge we faced with our mathematics team was the tension between content and instruction. Like other disciplinary teachers, there is real stress involved with covering so many units and topics during a semester or school year. The notion that teachers would have to explicitly include literacy skills in their instruction, potentially taking time away from content, met with understandable resistance. "Am I supposed to be a reading and writing teacher, as well as a mathematics teacher?" we sometimes heard. Because we understood this concern, our challenge was to help our teachers recognize the long-term value of incorporating literacy-based strategies alongside disciplinary instruction to simultaneously expand students' content knowledge and skill development. We did not want our teachers to think that teaching literacy skills would become a nuisance or that the effort wouldn't improve students' learning of mathematics content.

We recognize these concerns from the start because we want you to understand that literacy and mathematics instruction merge directly with the three big ideas of a PLC: (1) We believe in integrating change that is *focused on every student's learning*, where teams systematically consider, implement, evaluate, and revise all changes; (2) we believe in supporting *collaboration* (a collective commitment) between experts seeking to solve an educational concern; and (3) we believe that we must focus on the *results* students produce and use that knowledge to adjust and improve instruction (DuFour, DuFour, Eaker, Many, Mattos, & Muhammad, 2024).

We dedicate this book to the literacy issues we think warrant teachers' attention when connecting literacy to mathematics. We hope these ideas can help develop collaborative partnerships at your school, and we hope this book can serve as a strong resource for your teaching. Every teacher is a teacher of literacy.

INTRODUCTION

Every Teacher Is a Literacy Teacher

This book is part of the *Every Teacher Is a Literacy Teacher* series, which provides guidance on literacy-focused instruction and classroom strategies for grades preK–12. The elementary segment of this series includes separate titles focused on instruction in grades preK–1, grades 2–3, and grades 4–5. While each of these books follows a similar approach and structure, the content and examples they include address the discrete demands of each grade-level band. The secondary-level books we've crafted for this series focus on how subject-area teachers in grades 6–12 need specific instructional strategies to approach literacy in varying and innovative ways. To address this need, we designed each secondary-level book to do the following.

- Recognize the role every teacher must play in supporting the literacy development of students in all subject areas throughout their schooling.
- Provide commonly shared approaches to literacy that can help students develop stronger, more skillful habits of learning.
- Demonstrate how teachers can and should adapt literacy skills to support specific subject areas.
- Model how commitment to a PLC culture can support the innovative collaboration necessary to support the literacy growth and success of every student.
- Focus on creating literacy-based strategies in ways that promote the development of students' critical-thinking skills in each academic area.

You may immediately recognize how this approach differs from many traditional school practices and formats, where educators view literacy development as the job of English language arts (ELA) teachers, reading teachers, or teachers of English learners. It is often accepted practice that these teachers bear the responsibility of teaching skills such as vocabulary development, inferential skills, and writing. In many PLCs we have worked in, we've seen collaboration generally begin by teaming teachers within like disciplines. Mathematics teachers team with other mathematics teachers, science teachers team with other science teachers, social studies teachers team with other social studies teachers, and so on. When teams form according to discipline, they tend to focus only on their content and discipline-based skills. Although these traditional approaches may be effective, we recognize the need for schools to support changes that make teaching literacy a responsibility for all teachers. We propose that schools use the power of their PLC to adopt the collective commitment that every teacher is a literacy teacher. Mathematics teachers need to be literacy teachers. Science teachers need to be literacy teachers. Social studies teachers need to be literacy teachers. World language and fine arts teachers need to be literacy teachers.

In this book, we build our case for the importance of literacy instruction in all content areas and emphasize how building collaboration among middle school and high school mathematics teachers and literacy experts (specialists and coaches), in particular, is an impactful catalyst for supporting student growth. We also stress a strong commitment toward building instructional improvements that support the growth of every learner. When discipline-based teachers and literacy experts team up to connect literacy-based strategies with discipline-specific, subject-area instruction, they provide structure to both the content and the processes that students need to master. These teachers build stronger approaches to instruction that improve learning outcomes (Stephens et al., 2011).

We recognize that many schools do not have dedicated literacy experts available to collaborate with teachers on the challenges of building stronger readers and writers in the mathematics domain. To that end, we encourage you to use this book as a thought partner with your team or as your own personal literacy expert that can help you generate changes to support student learning. It will be a helpful companion as you deepen conversations and navigate choices that will positively affect student growth and development. It describes and gives examples of more than a dozen literacy-based strategies that you can integrate into the everyday mathematics classroom to guide your students toward higher literacy skills, which will guide your students toward mathematics literacy. You can use many of the strategies immediately; others require more preparation. In either case, we urge you to get started.

There are many reasons why mathematics teachers in grades 6–12 need to be literacy teachers. Lauren Tecca, Joanna Weaver, and John Chen (2023) find that integrating literacy strategies in mathematics classrooms initiates and promotes positive gains in mathematics assessments, increased ability to write and explain mathematics reasoning, and the ability to use mathematics vocabulary to communicate mathematical concepts. These skills correlate to the following literacy skills in the Common Core State Standards (CCSS) for Mathematics Standard Mathematical Practices (National Governors Association Center for Best Practices [NGA] & Council of Chief State School Officers [CCSSO], 2010):

1. Make sense of problems and persevere in solving them.
2. Reason abstractly and quantitatively.
3. Construct viable arguments and critique the reasoning of others.
4. Model with mathematics.
5. Use appropriate tools strategically.
6. Attend to precision.
7. Look for and make use of structure.
8. Look for and express regularity in repeated reasoning. (pp. 6–8)

In addition, reading about mathematical processes, writing about mathematics, and thinking like a mathematics expert require a mindset that focuses on elements of reading and writing that are fundamentally different from reading for pleasure.

Reading and writing about mathematics require the following:

- A close attention to detail and cause-and-effect relationships
- The ability to make estimates and predictions
- The ability to interpret symbols with fluency and understand patterns and trends
- The ability to draw an inference by relying on background knowledge and fact-based evidence to develop new true understandings

Literacy strategies create an infrastructure of support that allow students to learn independently and confidently meet these criteria. Mathematics teachers who focus on building stronger literacy strategies in their classrooms provide the necessary skills that support students' abilities to develop their own understandings and practical applications and extensions of the content. Given that students have such a wide range of vocabulary, reading levels, and a diverse range of background knowledge related to numeracy, it is all the more important to be successful in this endeavor. It requires your collaborative team to provide more literacy instruction along with the important mathematics content that is part of your curriculum.

This introduction begins the journey by exploring the need for literacy instruction, establishing an understanding of disciplinary literacy in mathematics, and detailing the scope of this book.

The Need for Literacy Instruction in Mathematics

Picture a student who is just beginning to learn how to read. What behaviors do you see as this student engages with text? What are they learning to do first? How are they grappling with the challenge of learning how to read? Chances are, you visualize this reader at the beginning stages, working to crack the alphabetic code—breaking apart and sounding out words, one syllable at a time, and likely dealing with simple language and colorful illustrations. The words the student is trying to read are already ones that they likely employ in conversation. This student is engaging in growing basic literacy skills: decoding, fluency, and automaticity. During this early phase of learning how to read, comprehension and meaning-making almost take a back seat to decoding. The reader is working on the mechanical process of learning to read.

As readers advance beyond the beginning stages of reading and grow in their abilities to read, they become more fully fluent and able to comprehend a text. They begin reading to learn rather than learning to read. At this point, the advanced reader possesses the ability to make meaning from what they read—the process of reading is no longer dedicated to the mechanical process of encoding and decoding a text. Instead, it is dedicated to learning and thinking. More advanced readers can infer from and analyze what they read in a book, as well as what they encounter in the physical world, even when they have limited experience with a topic. Such readers possess the critical literacy skills they will need to engage with mathematics in school and later in life.

Now, what about the reader who is somewhere between these two phases—the reader who is neither a beginner nor advanced? What about the student who can encode and decode but struggles to apply this information to form new understandings? The reality that all teachers know and experience in their classrooms is that many students fall into this place along the continuum, and many students leave high school without the essential life skill of being critically literate. In fact, the National Assessment of Educational Progress (NAEP) results detailed in *The Condition of Education 2020* (Hussar et al., 2020) suggest that only 33 percent of eighth-grade students (2019 data) and 37 percent of twelfth-grade students (2015 data) in the United States possess literacy skills at or above the level of proficiency, and over 65 percent have not met this readiness benchmark. Additionally, the percentage of students who tested below the NAEP basic literacy skill threshold grew for both fourth-, eighth-, and twelfth-grade students since the 2018 report (Hussar et al., 2020). This means that the literacy achievement gap is widening as a majority of students are moving through middle school and high school without developing the literacy skills necessary to be successful engaging in literacy-rich mathematics activities.

This is the group of students with whom we are most concerned in this book. We know that this large group of students requires more attention and a higher concentration on skill development. Moreover, a specific portion of these students will continue to need support in even basic literacy skill development. It is this portion of our student population that seems to be the conundrum—often, we find these are the students who teacher teams struggle to support.

Unfortunately, the literacy struggles many students experience are not always transparent, even though they are all too familiar in classrooms. The graph in figure I.1 (page 6) represents the increasing gap in literacy as students progress through their schooling, boldly demonstrating the challenges educators must work to solve. These students are in desperate need of instruction that cultivates the intermediate literacy skills that serve as a common foundation to disciplinary literacy. These skills include building academic vocabulary, self-monitoring comprehension, utilizing fix-it strategies to understand a text, and applying knowledge to a prompted task (Buehl, 2017). When all teachers acknowledge their responsibility to support literacy skills, collaborative teams can then shoulder the shared responsibility of student literacy and address these alarming statistics.

Research confirms a real need for disciplinary literacy instruction in the secondary classroom. Timothy Shanahan and Cynthia Shanahan (2008) note the following.

- Adolescents in the first quarter of the 21st century read no better—and perhaps worse—than the generations before them.
- For many students, the rate of growth toward college readiness actually decreases as students move from eighth to twelfth grade.

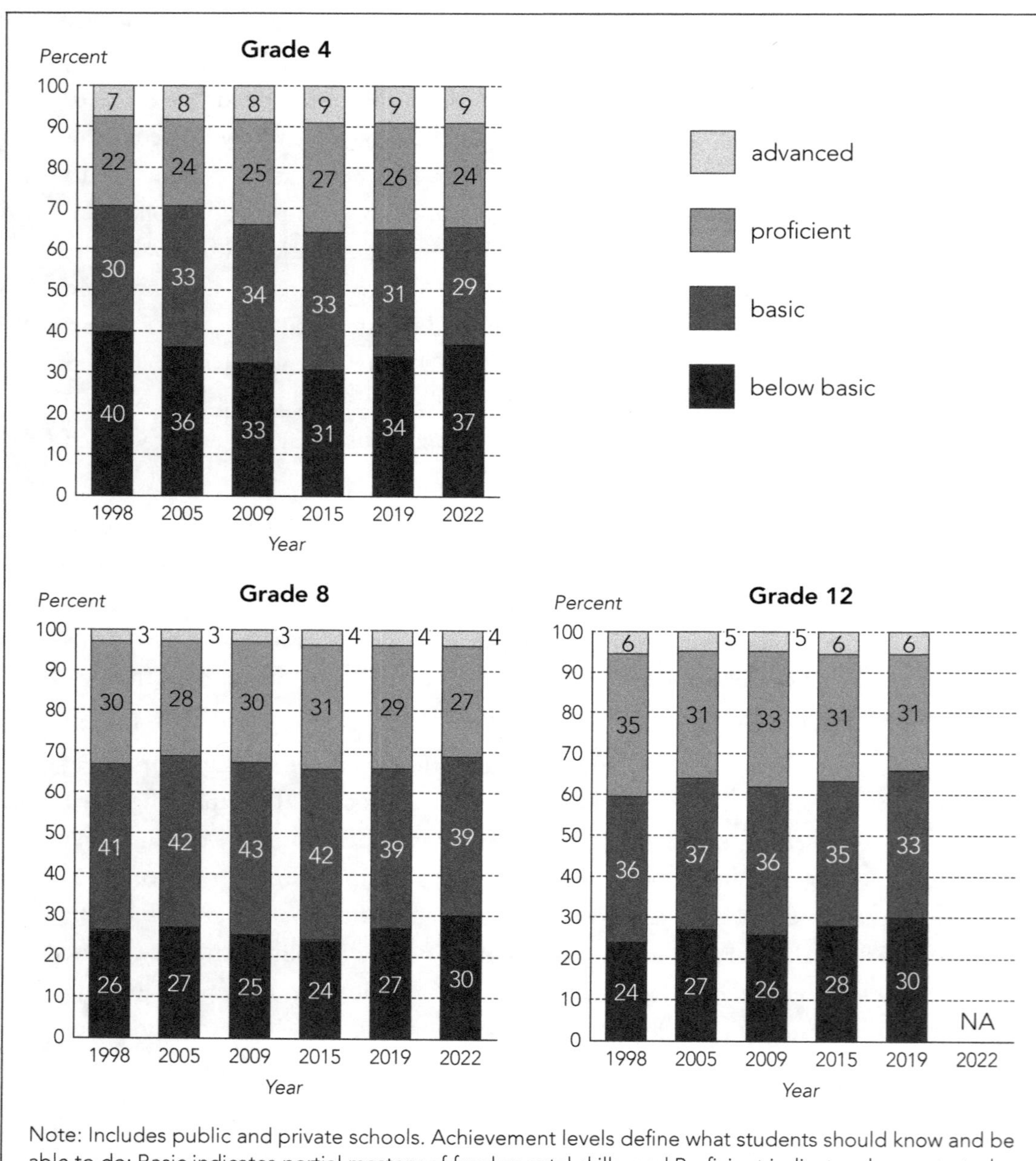

Note: Includes public and private schools. Achievement levels define what students should know and be able to do: Basic indicates partial mastery of fundamental skills, and Proficient indicates demonstrated competency over challenging subject matter. Although rounded numbers are displayed, the figures are based on unrounded estimates. Detail may not sum to totals because of rounding.

Source: U.S. Department of Education, National Center for Education Statistics, National Assessment of Educational Progress, n.d.

Figure I.1: Percentage distribution of fourth-, eighth-, and twelfth-grade students across NAEP reading achievement levels—selected years, 1998—2022.

- American fifteen-year-olds perform worse than their peers from fourteen other countries.
- Disciplinary literacy is an essential component of economic and social participation.
- Middle and high school students need ongoing literacy instruction because early childhood and elementary instruction do not correlate to later success.

Among the many concerns within collaborative discussions about teaching and learning, literacy continually ranks as one of the most worrisome. In many of our discussions with teachers across North America, teachers across academic disciplines express three common concerns: (1) many students struggle with basic literacy skills, (2) many students read and write below grade level, (3) and many students do not know how to accurately complete mathematical reading such as with word problems, write mathematically such as with geometry proofs, and explain themselves clearly when discussing mathematics. We believe that teachers can bridge these gaps with a strong foundation of high-impact literacy instruction and engaging in the PLC process.

The gaps we see in students' literacy skills are staggering, and these gaps affect all areas of many students' education. As students march through their schooling, the statistics demonstrate that gaps in literacy increase throughout many students' elementary, middle, and high school years. Columbia University Teachers College (2005) reports many students find themselves reading three to six grade levels below their peers, many students struggle mightily to comprehend informational texts, and many students graduate from high school unprepared to enter a college-level experience. Columbia University Teachers College (2005) and Michael A. Rebell (2008) further highlight the following statistics, which present significant and long-standing concerns.

- By age three, children of professionals have vocabularies that are nearly 50 percent greater than those of working-class children, and twice as large as those of children whose families are on welfare.
- By the end of fourth grade, Black, Hispanic, and poor students of all races are two years behind their wealthier, predominantly White peers in reading and mathematics. By eighth grade, they have slipped three years behind, and by twelfth grade, four years behind.

- Only one in fifty Hispanic and Black seventeen-year-olds can read and gain information from a specialized text (such as a STEM-focused journal) compared to about one in twelve White students.
- By the end of high school, Black and Hispanic students' reading and mathematics skills are roughly the same as those of White students in the eighth grade.
- Among eighteen- to twenty-four-year-olds, about 90 percent of Whites have either completed high school or earned a General Educational Development (GED) certificate. Among Blacks, the rate is 81 percent; among Hispanics, 63 percent.
- Black students are only about 50 percent as likely (and Hispanics about 33 percent as likely) as White students to earn a bachelor's degree by age twenty-nine, and elementary instruction does not correlate to later success.

In its U.S. Adult Literacy Facts infographic, ProLiteracy (n.d.) highlights the summation of this long-standing literacy crisis with the following facts detailing the reality of literacy in the United States and the catastrophic impact that illiteracy has on a multitude of social and economic factors. For example, more than thirty-six million U.S. adults cannot read, write, or do mathematics above a third-grade level. Forty-three percent of adults with low literacy levels live in poverty. When parents have low literacy levels, their children have a 72 percent chance of performing at the lowest reading level, receiving poor grades, developing behavior problems, having high school absentee problems, and dropping out. More than 1.2 million students drop out of high school each year (one in every six). These jarring statistics undoubtedly reveal a systematic divide between those who are literate and those who are not, consequently deepening the inequities already present in our social structures. Visit https://tinyurl.com/vefwshn3 to read all of ProLiteracy's U.S. Literacy Facts, including statistics regarding English learners, unemployment, health literacy, and correctional facilities.

Statistical results like these are a stark reminder that teachers need to focus their attention on the literacy and mathematical development of every student in their schools. As we will note throughout this book, reading and writing strategies in the mathematics classroom often require differing instructional approaches that demand innovative thinking. Since mathematics does not require traditional reading and writing tasks, strategies and lessons need to be adapted in order to fit the different types of literacy required for mathematics, such as word problems or proof writing. Plus, teachers must tailor approaches to meet the needs of every student.

In this book, we offer suggestions focused on teaching students intermediate literacy skills—skills needed to help students transition from early, foundational literacy to specialized disciplinary literacy. These are important skills to attain because students with strong intermediate literacy skills have essentially developed an awareness of their own active comprehension, and they know what to do when their comprehension begins to feel shaky. It is vital within all disciplines, including mathematics, that teachers don't jump ahead of intermediate literacy but instead continually model this phase for students and provide opportunities for them to practice these skills in a constructive and guided manner independently and confidently.

There is a focus on literacy throughout all of the CCSS (NGA & CCSSO, 2010). In this book, we regularly refer to the Common Core State Standards for Mathematics (NGA & CCSSO, 2010), specifically the Standards for Mathematical Practice that help to articulate the process standards as priorities teachers should teach and employ in support of students' understanding and mastery of mathematical content standards. In doing so, we strive to emphasize the importance for all students to have the literacy expertise necessary to be college- and career-ready after graduating. Additionally, when mathematics teacher teams and literacy experts collaborate on these standards, they are sure to make a powerful impact on student learning. It is important to note that even if some schools and states do not use CCSS in favor of another set of standards, the CCSS can still serve as a model of the role literacy plays in the discipline of mathematics.

Disciplinary Literacy

As your team gains confidence that students have a good grasp of basic, foundational literacy skills, and as team members begin to see students develop more intermediate and advanced literacy skills, your team can move forward with tailoring literacy instruction with an eye toward disciplinary literacy. Even though students will need you to continue modeling the use of academic vocabulary and monitoring their comprehension, they will also be ready to attack complex texts—word problems, geometry proofs, mathematics vocabulary, and more—with a disciplinary lens even as they practice building their skills. Mathematics teachers are experts in their respective content areas. We hope that the tools in this book will allow teachers to help students build their own reading in mathematics toolbox.

For our purposes, a *discipline* is a unique expertise that schools often split into subject matter divisions such as mathematics, science, English, physical education, world languages, fine arts, and so on. *Disciplinary literacy* focuses on the literacy strategies tailored to a particular academic subject area. This book focuses on the expertise of mathematics teachers who see the value of integrating specific literacy-building strategies into their classrooms because of the consistent employment of the complex literacy tasks mathematics texts require—often the first challenge students face prior to getting to the core content.

What would happen if we were to gather teachers from every discipline in a school and track the way they each address reading, writing, and speaking tasks? Predict how different content-area teachers would approach and work through literacy tasks. What similarities and differences would we observe among these varied disciplines?

Because all teachers have unique expertise related to their individual academic fields, they often approach literacy-based tasks differently. Those differences stem from the diverse sets of expertise, interests, and background knowledge each professional brings to teaching and learning, and as a result, middle school or high school teachers often attend to literacy tasks differently based on that disciplinary expertise. After all, when a mathematics teacher reads, writes, and speaks, they do so with certain goals and objectives in mind, such as determining what a problem is asking, identifying important details and variables, explaining mathematic principles, and explaining problem solving, to name a few. While this may be a similar goal to other disciplines, what is specific to mathematics is that the output of each of these skills represents and articulates the mathematical story at hand and a specific solution. As such, the stakes for comprehending mathematics literacy content are high.

There are also certain stylistic and conceptual norms professionals attend to in each discipline. A mathematician, scientist, historian, businessperson, or any other professional addresses literacy tasks with norms and behaviors befitting their expertise and profession. That makes total sense; after all, each expert or professional has unique insider knowledge, with more background knowledge, subject-related vocabulary, and subject-related purpose than others without such dispositions. As a result, disciplinary outsiders often lack sufficient background knowledge and vocabulary to navigate a disciplinary text successfully. Literacy expert Doug Buehl (2017) suggests that the job of educators is to teach students how to think like they do—as disciplinary insiders. A social studies insider might approach reading tasks with specific goals and objectives, such as comprehending text as a means to understand the past and to apply new understandings to the world around the reader, whereas an English insider might focus on themes and symbolism. A mathematics expert might turn a story into numbers and also solve and evolve those numbers into words, further developing and articulating the story behind the mathematics.

Text comprehension in all disciplines generally follows a similar nine-step process, as illustrated in figure I.2 (page 12). (See page 124 for a reproducible version of this process.) This figure and process derive from a guide Katherine Gillies (2019) developed when training peer tutors to help struggling readers navigate disciplinary texts; the figure provides a pathway for students to follow and teachers to model when comprehending a text. As we explore strategic comprehension steps and before, during, and after stages of reading, all related to mathematics in more detail throughout, we will demonstrate how the application, connection, and extension of literacy skills unfold under the specific disciplinary lens of mathematics.

Given the difference between disciplinary insiders and outsiders, it makes little sense for content-area teachers to instruct students to read and write with the same general strategies and moves in every content area. After all, if we know that each content area has its own thinking style, it makes sense that teachers support students in consuming and producing texts with the same unique thinking style required of each discipline. Even students who have a solid foundation of general strategies may struggle with the specific demands of disciplinary texts. Instead of using generic strategies in every class and across the school, providing students with a varied strategy toolbox to meet disciplinary demands better equips them as disciplinary insiders to read like mathematicians, scientists, historians, and so on (Gabriel & Wenz, 2017).

Over time, our literacy team has made positive strides toward building disciplinary literacy strategies that support learning in more directed, focused, and attentive ways. We've learned that we should apply more specific strategies to different disciplines in ways that help support learning. When we speak of this shift to disciplinary literacy and training students to be insiders, we intend to teach students to think differently in each classroom they encounter during their day. We encourage schools to establish a common literacy vocabulary for teachers to use across the school. We believe this helps students to build a general understanding of common terms such as *claim*, *evidence*, and *reasoning*. Through this common use of language, students can build on the foundational knowledge of these literacy concepts in discipline-specific ways as teachers become more comfortable teaching

<table>
<tr><th>Did I . . . ?</th><th>Strategic Comprehension Step</th><th>Before, During, or After Reading</th></tr>
<tr><td>☐</td><td>Preview text, ask questions, and make predictions.</td><td rowspan="5">Before:
Focus and get ready to read.</td></tr>
<tr><td>☐</td><td>Recall what you already know about the topic.</td></tr>
<tr><td>☐</td><td>Set a purpose for reading.</td></tr>
<tr><td>☐</td><td>Make a notetaking plan for remembering what's important.</td></tr>
<tr><td>☐</td><td>Define key concepts and important vocabulary whenever possible.</td></tr>
<tr><td>☐</td><td>Keep your purpose for reading in mind.</td><td rowspan="3">During:
Stay mentally active.</td></tr>
<tr><td>☐</td><td>Make meaning by:
• Asking questions
• Putting the main ideas into your own words
• Visualizing what you read
• Making notes to remember what's important
• Making connections between the text and people, places, things, or ideas</td></tr>
<tr><td>☐</td><td>Be aware of what's happening in your mind as you read. Consider:
• Am I focused or distracted?
• Do I need to go back to a part I didn't get and reread it?
• What are my reactions to what I am reading?</td></tr>
<tr><td>☐</td><td>Reflect on what you've read. Consider:
• Did I find out what I needed or wanted to know?
• Can I summarize the main ideas and important details in my own words?
• Can I apply what I have learned?
• Can I talk about or write about what I have learned?</td><td>After:
Check for understanding.</td></tr>
</table>

Source: © 2019 Katherine Gillies. Adapted with permission.

Figure I.2: Reading-comprehension process poster.

Visit ***go.SolutionTree.com/literacy*** *for a free reproducible version of this figure.*

students how to read, write, and think like experts in their classrooms. This is the goal of disciplinary literacy and why we often ask teachers who wonder how to teach a text, "How would you, as an expert, address the task?" As they think through their own processes, often a strategy or a focus emerges that is unique to their discipline. This allows us to help teachers recognize the value of thinking about their discipline in relation to literacy, which is particularly important in the teaching of mathematics, as students work to gain insights into the texts they read, weigh and value a text's details, and develop their capacity to convey a text with numbers, problem solve, and then articulate that story using words while critically analyzing problems and information.

About This Book

Our goal for this book is to support collaborative partnerships in schools to better equip mathematics experts with methods that enhance what they are already doing in their classrooms and that better equip them to respond to students' literacy needs. We see literacy as part of the equation to build more successful mathematics students. We aim to strengthen teams' use of literacy strategies to help all their students develop skills as readers and problem solvers. Collaboration on these strategies will create new ways to heighten students' abilities to approach more complex texts with confidence and advance their abilities to think critically. The following sections explore the scope of this book, the common language we use throughout, and the individual chapter contents.

Scope

We designed this book to help literacy leaders, content-area teachers, and school leaders collaborate and build literacy capacities in secondary-school environments. We talk more specifically about what makes a literacy leader in the next section, but what's important to know here is that anyone in your building who takes on the mantle for driving literacy advancement in the classroom is a literacy leader. In elementary school, teachers work hard to teach students to *learn to read*. In middle school and high school, the goal is to teach students to *read to learn*. There is a big difference between the two approaches, as we noted earlier (see The Need for Literacy Instruction in Mathematics, page 4). Moreover, mathematics teachers want students to learn from the literacy tasks, such as a graphic organizer that requires them to dissect a problem, that prompt the critical thinking about the reading of problems and the solutions they write, moving toward metacognition.

As your team works to approach these challenges, team members need to recognize that each school is unique and each student is unique: there is no one-size-fits-all pathway to literacy development. Within this book, there is a continuum of supports related to the varying needs of each school and each classroom. Sometimes, teachers might require short-term, immediate literacy triage; sometimes long-term, sustained collaborative development between team members is necessary; or sometimes there is a need for *both* triage and sustained literacy-based professional development. We recognize that strong, consistently applied literacy strategies can and will help all readers develop their potential, so we invite you to adapt the strategies we offer in this book to your unique needs. Many of the same literacy strategies that work for less complex literacy tasks still apply to more complex tasks—the only difference is the difficulty level. The skills that students need to apply remain the same and, with consistent application, become ingrained habits of the mind. As mathematics teams collaborate on their work, staying committed to literacy-based strategies will help all students advance.

Common Language

For the purposes of this book, and to avoid getting confused by education jargon, we recognize that we need to have a common understanding of literacy and a common language about literacy development. For instance, we use the word *text* to mean a textbook reading, a mathematics problem, an article, a chart, a diagram, a proof, a cartoon, a media artifact, and so on. There are many texts we ask students to read and interpret, and they can be in many formats. In addition, the term *literacy leader* can apply to a variety of educational roles. Throughout the book, a literacy leader can be anyone in your building, such as an administrator, teacher leader, reading specialist, or literacy coach. A literacy leader is someone who has a knowledge base in literacy and wants to improve the overall literacy skills of a school environment or institution. If you don't have a literacy leader at your school, don't let that stop you. Remember, you can use this book as a thought partner and become your school's literacy leader. The goal is to get started with the demanding challenges of literacy that need to be tackled now, with or without a literacy coach or a preexisting school literacy leader championing the work. Any teacher and team of teachers can initiate the changes that are necessary to support student learning; we mean for this book to help guide your understanding of how to approach these changes in teaching practices.

In this book, you will frequently see the term *professional learning community (PLC)*. A *PLC* is "an ongoing process in which educators work collaboratively in recurring cycles of collective inquiry and action research to achieve better results for the students they serve" (DuFouret al., 2024, p. 14). For example, in our PLC, teams often look at student artifacts, performance, and data in order to make instructional decisions that affect our current and future instruction. As we established in the preface to this book (page xv), a PLC consists of a schoolwide or districtwide culture of *collaboration* that uses a *results orientation* to achieve *learning for all*. We believe that a commitment to these three big ideas can help to support and innovate literacy in every classroom, and we believe that PLC culture promotes changes that will effectively support all students.

Within a PLC culture, collaborative teams meet on a consistent basis to build innovative practices concentrated on student growth and learning. We use the term *team* throughout the book with the understanding that all teams are interdependent and are professionally committed to continuous improvement. We know that teams may look different from building to building, and we know that schools need to configure teams differently based on building resources. In this book, we use *teams* generically to refer to teachers who are collaborating in focused ways to address literacy concerns for student learning in their classrooms. We also recognize that you might be a team of one—a singleton instructor who teaches an elective course within the mathematics division or the only person teaching a grade-level course. For those singletons, we encourage you to be creative in finding ways to collaborate and discuss how to make use of literacy strategies more effectively. Reach out to teachers of different grade levels or in other area schools. Or use the community forum at the AllThingsPLC website (www.allthingsplc.com) to locate a content-area teacher to collaborate with. There is great value in collaborating about how to use a strategy or make it more effective for your specific students, and in a PLC, every educator must be part of a collaborative team.

Chapter Contents

In chapter 1, we lay out the fundamental aspects of collaborative teams and the processes they engage in to address teaching literacy in the mathematics classroom. In chapter 2, we begin with more in-depth discussions about foundational literacy and many immediate interventions for literacy difficulties that require a fast solution. We call this *literacy triage*. In chapter 3, we address the importance of thinking like a mathematician. From there, we focus on disciplinary literacy collaboration for prereading, during reading, and postreading in chapters 4 and 5. Chapter 6 then offers guidance for teaching writing in the mathematics classroom. Within these chapters of the book, we slow down intentionally to support a deeper, focused approach. We offer classroom strategies that are the result of collaborative explorations by literacy leaders and content-area teachers, providing clarity on how varying perspectives inform instruction within the mathematics classroom. For each example, we discuss that particular strategy's purpose, application within mathematics, and literacy focus. We also indicate how teams can modify each strategy to support students who are struggling (including students who qualify for special education or are learning English) or to extend learning for students showing proficiency. Even though these adaptations are geared toward these subgroups, they apply to any students who would benefit from differentiation. Finally, chapter 7 covers ideas for formative and summative assessment and feedback.

Throughout this text, there are opportunities called Thinking Breaks. We intend for these to help you reflect on current practices, challenges, and opportunities for growth in working with literacy in the mathematics classroom. We know that you might do this naturally, but these are the points where we think it is important to slow down and consider ways to apply the strategies we suggest to your own students. In addition, each chapter ends with a series of questions to encourage thinking about your team's collaborative considerations. These questions highlight ways for teams to discuss, collaborate on, or implement disciplinary literacy ideas. Your team can use these discussions to build more directed literacy practices as you target your specific grade-level curriculum.

Ultimately, we hope this book and series are not only resources for ideas you can implement immediately in the classroom but also sources of inspiration for collaborative opportunities between literacy leaders, mathematics teachers, and school leaders to increase literacy capacity in your building or buildings. As a mathematics teacher, your content provides a unique opportunity for students to engage critically with text on a regular and consistent basis, creating a natural opportunity for students to further develop their literacy skills.

As I am reading and using this book as a resource to support my teaching, what do I want to get out of the content?

Note these three considerations for your team: (1) use this book as a book study, (2) break the book down chapter by chapter and focus on specific changes, and (3) prioritize your concerns for student learning and how to best support the literacy development of your mathematics students.

Wrapping Up

Building collaborative teams focused on literacy development can be challenging. We know you are extremely busy and have enormous amounts of content to cover, so you may be reluctant to add another layer to your already demanding workload. However, given the NAEP data showing that more than half of U.S. twelfth graders graduate high school without preparation for advanced critical thinking (Hussar et al., 2020), we must pause and consider what we are all doing as educators to better prepare students for the future. Providing students with important intermediate literacy and disciplinary literacy skills is an important step toward building the advanced literacy proficiency they will need throughout their educational careers and beyond.

Collaborative Considerations *for* Teams

- What are some of the unique features of texts teachers use in mathematics classrooms?
- What kind of criteria do experts in your discipline look for when they read?
- What deficiencies do you notice in your students that might obstruct their understanding of your content?
- How might your team provide experiences and vocabulary to help students feel more confident in reading texts used in a mathematics classroom?

CHAPTER 1

Collaboration, Learning, and Results

It is no secret that educators have a lot on their plates, so convincing content-area teachers that literacy is valuable is only one step on the way to improving literacy at a school. We all need to teach literacy to the benefit of our students; the natural next step is finding time and building collaboration. Many are willing to teach literacy but do not know how—and all are very busy, which makes finding the time to prioritize this work a challenge. Just like any other content-area team, a collaborative mathematics team needs to ask difficult questions about its curriculum and teaching practices, even ones that question the practices that the team has believed are tried and true methods or curriculum components. Identifying common concerns as a department almost always leads to an opportunity to determine a solution-focused literacy need, like strengthening students' academics and content-specific vocabulary knowledge, building student capacity to apply this vocabulary knowledge to the instructions, and consequently, the narrative behind mathematical thought and process. Here is your opportunity to begin the collaborative process. Is it possible that a focus on literacy strategies for mathematical problem solving might be what your students actually need to be more successful? Maybe some of the suggestions in this book will offer you and your team some out-of-the-box ideas to help revise or enhance your instruction. We hope so.

For us, collaboration plays a crucial role in the success of any school dedicated to building effective teams in a PLC. When experts collaborate, innovative ideas emerge in ways that support student learning and generate positive results. Around the world, forward-thinking schools are making a commitment to this core principle, they are tackling long-standing concerns in education by bringing together teacher teams to make stronger curricular and instructional choices, and they are getting better and better at making use of assessment practices that support the formative development of all students.

This chapter helps you identify how to initiate collaboration around literacy by applying PLC fundamentals and building teacher teams within your school to support meaningful teamwork that leads to student growth and reflective teaching practices. We offer guidance for mathematics teams and examine how to approach meeting logistics. We then examine in detail the work teams carry out in collaborative meetings, including analyzing standards, setting goals, identifying students' existing literacy skills and needs, and finding connections between mathematics and literacy skills.

How It Works (Who Exactly Teaches Literacy Skills?)

Whose job is it to teach literacy skills to our students, from those who are struggling to students who are excelling? Some may assume it is the job of the English department or the small reading department that may exist in some schools. While there is truth to the statement that these teachers must take responsibility for explicitly teaching students how to navigate text and provide scaffolds and supports so that students' literacy abilities will grow, this is also true for teachers of all content areas. This means that we cannot assume that the English department is the sole responsible party for the cultivator of student literacy growth. In fact, in reality, while reading teachers are skilled in methods of research-based instructional intervention and support, one class period a day will in no way be enough to stretch and grow student skills to where they need to be post–high school; all disciplinary teachers must get in on the game and be active players in supporting student literacy growth.

On the surface, it may seem like a relatively easy task to get your mathematics department colleagues on board with the idea of supporting student literacy growth because the content is so text-rich; however, remember that while some may be skilled in teaching literacy skills naturally, this is not something that was covered extensively in most teacher-preparation programs. Just like any other disciplinary team, a mathematics literacy team will need to ask difficult questions of curriculum and teaching practices. Mathematics teachers are a busy bunch—they are required to cover a lot of mathematics concepts in a single course, which can be completely overwhelming, and they can easily fall into the role of summarizer or formula provider for students instead of asking students to struggle through reading their textbooks and problems. However, identifying common concerns as a department almost always leads to an opportunity to identify a solution-focused literacy need—we invite you to begin this important collaboration process.

Collaboration in a PLC

Collaborative teams are a pivotal force of progress in PLCs. While the size and scope of work in teams can differ greatly, the initial focus of a team often starts with a specific, discrete task that later evolves into more layered tasks and discussion topics. By simple description, a *professional learning community* is a group of professionals who come together to collaborate; more specifically though, the team members are focused on three big ideas: (1) a focus on learning, (2) a collaborative culture, and (3) a focus on results (DuFour et al., 2024). Within our literacy work with disciplinary teachers, we keep these three big ideas at the core, and we direct our commitment to literacy in all disciplines by continuously addressing the four critical questions of a PLC.

1. What knowledge, skills, and dispositions should every student acquire as a result of this unit, this course, or this grade level?
2. How will we know when each student has acquired the essential knowledge and skills?
3. How will we respond when some students do not learn?
4. How will we extend learning for students who are already proficient? (DuFour et al., 2024, p. 44)

We recognize that teams are configured in varying ways depending on the school. For instance, you might have curriculum teams, grade-level teams, content-focused teams, or teams comprised of singletons (the lone teacher for a grade level or content area in a school). No matter how your teams are currently structured, when working toward integrating literacy-based strategies, we hope they have the means to collaborate with a literacy expert in your building—an expert who can provide insight into varying and supportive literacy strategies for teachers to integrate into their instructional practices. As we wrote in the introduction, though, if your building does not have a dedicated literacy expert, we designed this book also to function as a thought-provoking substitute that will guide you in generating ideas and making actionable plans.

While the size and scope of work in a collaborative team can differ greatly from one context to another, many teams' initial focus is often a specific, discrete task that later evolves into more layered tasks and discussion topics. A *team* is a group of professionals who come together to collaborate. PLC architect Richard DuFour (2004; DuFour et al., 2024) notes the following about successful teams.

- Have common time for collaboration on a regular basis.
- Build buy-in toward a discrete and overarching common goal.
- Build a sense of community.
- Engage in long-term work that continues from year to year.
- Grow—but do not completely change—membership each year.

In addition to these characteristics, in our experience, it is helpful to have a team leader or point person who creates agendas and monitors discussion (this can be a rotating role); it is important to encourage open, honest discussion-based dialogue that respects and includes all ideas concerning student learning. We recommend teams find opportunities to reach out and include any colleagues whose work might overlap with the team's. This might include a reading specialist, a teacher of English learners, or a literacy coach; additionally, you might reach out to other disciplinary teachers to identify overlapping areas of literacy instruction. The following sections go into more detail on the configuration of teams, the role of leaders within a team, and the logistics for team meetings.

Team Configuration

No single team organizational model is essential to this kind of focused collaboration. For some schools, you may already have grade-level teams in place within your department, or you may have vertical teams composed of representation from each grade-level curriculum. Ultimately, how your team is structured will drive your goals. Whether your team is vertical or horizontal, you will need to consider the endgame goal for your students—whether that is a multiyear curriculum (at the top of the stairs) or at the end of the current school year (still working on climbing the curriculum staircase).

At this point, we want to make a few explicit recommendations for configuring teams effectively when working on mathematics literacy-based strategies. There are a number of different approaches to establishing a team, and in many schools, resources vary. For instance, many schools might not have dedicated literacy experts trained in the value and capacity of literacy strategies. Sometimes, we need to think of teams differently if we are thinking about how to make changes happen. Here are a few considerations when building teams focused on literacy.

1. Ideally, incorporate a literacy expert on a curriculum- or content-focused mathematics team to serve as an informed thought partner. An individual with literacy expertise can help support instructional changes with a greater variety of approaches and can also help select strategies that are better aligned with course outcomes. For purposes of the mathematics classroom, the literacy expert can help to adapt strategies that are focused on decoding the text or problems, summarizing the problems or the answers, making inferences, or text extensions and connections, and helping students learn how to explain their thinking as they problem solve instead of just trying to get the right answer.
2. Some teams work *horizontally*, meaning members work within their specific grade level or within a particular content-based course with many sections and many teachers. Some teams work *vertically*, where they meet with teachers from different grade levels to ensure curriculums are interconnected and work to build learning year to year. When considering literacy-based strategies, think about how your team is constructed and the purposes of your goals for integrating literacy-based strategies. Why and how should your team apply strategies to support students throughout a particular grade level or within a mathematics course that many teachers instruct? When and why should a team use a strategy? If working vertically, how and why might team members teach and reteach strategies year after year while ensuring students reach the next course or grade level with the essential knowledge they will need? How do these strategies become habits over the sequence of courses and throughout grade levels?
3. Some teams will not include or have access to a literacy expert. In these cases, consider what other resources can serve as a strong, literacy-informed, and reflective collaborative partner. Consider the expertise that might come from other academic departments, such as ELA, where there may already be well-established practices for literacy instruction. Is there a teacher with experience coteaching alongside a reading teacher who could be available to come to your team meetings? If not in your building, is there a literacy expert in a building within your district—an expert in an elementary, middle, or high school building who might be able to help?

Is there expertise in the special education department or an intervention specialist with a background in the area of literacy? Seek to use the resources available to you.

4. Consider how your team might make collaborative use of this book in more directed ways. In the absence of a literacy specialist or coach in your school, we hope that this text will provide you with accessible approaches to work with your colleagues to impact literacy instruction in the mathematics classroom. To improve, teams often need to seek outside resources that do not currently exist within the schools they are working in. If you are a singleton (one teacher) or on a smaller course team, widen your definition of a team and consider the larger community of educators that is willing and ready to help other teachers learn. Reach out to professional organizations, attend conferences, or use tools at AllThingPLC (www.allthingsPLC.com) that can help connect your team to discussions that will influence positive changes.

It is important to note that regardless of your team structure, the only real participation requirement is having people who are willing to problem solve and work together. We have been part of teams that developed organically simply because a group of teachers had a common concern or question, as well as those that were mandated as part of schoolwide improvement plans—the same outcome can be yielded with careful planning, honest rapport building, and a team approach to inquiry to build buy-in and promote investment.

Leaders' Roles

Collaboration is at the crux of literacy work and is truly an essential component of professional growth. While this may seem obvious, achieving authentic collaboration can be a challenge for several reasons. As discussed previously, we commonly hear "There isn't time" as a primary roadblock, since mathematics teachers, like all teachers, feel pressure to cover so much material with their students. Consequently, the first step in the collaboration process takes place between literacy leaders and administrators who need to work together to find time to build effective, supportive changes to create a long-term plan for literacy teaching and learning in the mathematics classroom.

Your past experiences with using immediate strategies to address students' most critical literacy needs provide an opportunity to approach your administrators and create a long-term plan for schoolwide literacy. The following chapters provide information and examples of how to begin this process, but here are some starting points to consider when communicating with administration and when engaging with the chapters that follow in this book.

- Identify the immediate problem and show the evidence of the problem with data. Is there assessment data that indicates students have gaps in mathematical literacy skills? Consider surveying your students to identify their confidence or efficacy in applying reading and writing skills with mathematics-based tasks. Be able to specifically name where there are literacy gaps and how those are emerging in student learning.
- Identify your ultimate goal for the students in your classroom. Know what you want to change. As a team, brainstorm ideas of what literacy success looks like in your content area. By establishing a clear set of aspirations for student success, the team will have a clearer sense of what they want students to learn in order to close literacy gaps.
- Provide a list of strategies you have tried in your classroom. Identify the ways you and your team have worked to build change.
- Provide suggestions for resources or ideas that can help you and your team accomplish your goals and get your students to a more successful place. Collaborate with your administrators on how your team can accomplish these goals, with the needed action steps and the right supports in place.

Responding to your school's literacy needs cannot occur through an occasional meeting or a purchased program. Everyone needs to be on board with the challenging work of moving students' literacy competencies in the right direction, and leadership must support and promote this collaboration to ensure each teacher is open to their own education and professional growth. Having said that, we want to offer a few tips for literacy leaders engaged in this work with mathematics professionals.

If you are the literacy leader in your school, consider yourself a host to other teaching teams. You want this team collaboration with you to be as positive as possible so everyone can be an effective participant. Often as a leader in literacy, you will need to serve as the glue that holds the pieces together. You will need to develop the agendas and send out the invites and reminders. Your personal goal is to keep this team moving forward and committed to literacy-based strategies. Be authentic: you are not the person with all the answers; you really only have half the answers. The key to collaborating with mathematics experts about literacy is to effectively connect literacy strategies with the content and skills in mathematics. Remember that as mathematics teachers, you are the experts. How can we collectively demystify what it means to "think like a mathematician" for our students so they have the tools to be successful? The good work that emerges will be the result of innovative, thoughtful collaboration between the literacy expert and the mathematics experts—not as the result of any one individual leader.

What additional questions or thoughts could you consider when approaching administrators about creating a schoolwide literacy program that can specifically assist in the mathematics classroom?

Collaborative Meeting Logistics

Based on our experiences, there are four simple logistical action steps you can take that will help ensure fluid and timely collaboration.

1. **Work creatively and collaboratively to carve out a common regular time and space for your team to meet:** From our experiences, carving out common time is the first step to developing a high-functioning team in a PLC. While a progressive education model allows for consistent and regular professional collaboration during the school day (due to the complex nature of the secondary school day), many schools do not have a schedule that reflects this. And while a regular weekly team block where students start school late or leave school early is often the ideal way to ensure teacher teams can meet in focused ways, we also recognize that many school districts are still working toward supporting structures that empower the PLC culture. If your school has not yet set up a structured time for teacher teams to meet, we strongly suggest planning ahead to ensure team meetings allow for collaboration about literacy. If you and your teacher colleagues are dedicated to literacy in your classrooms, you might need to work with an administrator who will help support the collaborative time necessary to innovate positive changes. Ask for the time you will need to collaborate and innovate literacy strategies to use in your classrooms. For example, you might ask for release periods or release days throughout the year so you can accomplish your team's goals.

2. **Create a regular meeting schedule, and make sure everyone knows the plan:** Collaborative teams must ensure they dedicate their meeting time to fostering the commitments of a continuously developing PLC. Early on, establishing norms that set focused, actionable goals helps teams achieve their purpose and helps to establish commitments. We suggest using SMART (specific, measurable, actionable, results oriented, and time bound) goals as a good guiding tool (Conzemius & O'Neill, 2014). By setting up SMART goals, your team is more likely to stay focused on learning and driven to succeed. As literacy coaches, we know that time is precious, and when our mathematics team teachers meet with us, we know that our collaboration needs to have purposeful, specific outcomes. In building SMART goals toward literacy, we encourage taking the time to create an action-driven schedule that is paced, practical, and respectful of everyone's many different commitments. What is realistic depends on your structure and your team's purpose. From our personal experiences, we think that meeting less than every other week often means that team members will be unable to prioritize making changes in literacy practices. Team meetings that are too infrequent mainly consist of recapping what happened at the last meeting and miss the mark on cultivating the productive collaboration that leads to changes to literacy practices in the classroom. To ensure all team members are aware of upcoming meetings, create electronic calendar invites, send out a paper copy of dates and plans that members can post at their desks, and send reminders. Always attach your agenda to encourage thoughtful preparation prior to the meeting.
3. **Identify and use a consistent meeting location:** We encourage finding a consistent space for your team to meet. It is counterproductive when people are always searching for a changing location. Ideally, this space will be free of other distractions, comfortable, and well-equipped for the work you will be doing (for example, you have access to a whiteboard, projector, laptop, and so on).
4. **Create digital files that capture agendas and notes:** From the beginning of your work, create an ongoing digital hub for agendas and notes—use tools like Slack, Google Docs with Google Drive, or Microsoft OneNote with Microsoft OneDrive. Ideally, you want something that allows all members to contribute independently and ensure full, online access to your materials outside of your collaboration time.

thinking BREAK

- Do you see any natural opportunities within your personal schedule for team time? Do you have a prep period in common with other colleagues?
- Are there any predictable patterns you notice in your department or school's master schedule that would allow for meeting time? Does your school have an altered bell schedule periodically for collaboration or other types of professional work?

Standards Analysis and Goal Setting

As we noted before, collaborative teams in a PLC agree on what they want all students to learn, thus addressing the first critical question of a PLC: What do we want students to know and be able to do? (DuFour et al., 2024). While your team may have a general overarching literacy goal right out of the gate, it must still dedicate time to the specific development of discrete goals that identify what students are to learn. Often, it is necessary to do some research to establish an effective goal. Before defining any sort of common goal for student learning, spend time examining your school's mathematics curriculum standards, as well as the current state of implementation of these standards, with a productive and critical eye. Your team must unpack the mathematics content standards and identify the literacy skills they require for mastery. Once unpacked, set aside the complete content standards momentarily to focus purely on the skills that serve as the vehicle to transport the learner from novice to skilled within the content mastery continuum.

Use figure 1.1 (page 29) to discuss with your team the content standards you are prioritizing and the evidence of student learning that you expect students to demonstrate. This process ensures your team addresses the first two critical questions of a PLC: What do we want students to know and be able to do, and how do we know when they have learned it? (DuFour et al., 2024).

We urge you to not overlook the steps in the reproducible "Content-Standard Analysis Tool" (page 125). While it is true that content standards may shift over time, it is also critical that we continue to evolve curriculum alongside changing standards. While it is tempting to jump to identifying power standards, make sure that you don't skip the unpacking of *all* standards first. Curriculum teams that overlook nonpower standards often fall into the trap of simply overstating core curriculum components—essentially saying, "We do all of this already," and treating

<table>
<tr><th rowspan="2">Content Standard</th><th colspan="2">Artifacts</th><th rowspan="2">Power Standard
☐</th></tr>
<tr><th>Lesson Plans and Course Materials</th><th>Student-Generated Evidence of Standard</th></tr>
<tr><td rowspan="3">Unpacked (process standards):</td><td></td><td></td><td rowspan="3">Opportunities and Next Steps</td></tr>
<tr><td>Strengths and Needs</td><td>Strengths and Needs</td></tr>
<tr><td></td><td></td></tr>
</table>

Content Standard-Analysis Steps (Use this with your team to guide collaborative discussions.)

1. **Unpack:** What is the standard really saying? Put the standard in your own words and determine the individual process standards inherent within the content standard.
2. **Identify artifacts:**
   - Lesson plans and course materials—
     - How is the standard being taught?
   - Student-generated evidence—
     - How are students demonstrating mastery?
3. **Assess the strengths and weaknesses of these artifacts:** Where is there room for improvement?
4. **Identify power standards:** Which content outcomes are the most essential to your content area?
5. **Identify next steps:** As your team's work continues, identify potential teaching and learning opportunities.

Figure 1.1: Content-standard analysis tool.

Visit ***go.SolutionTree.com/literacy*** *for a free reproducible version of this figure.*

a standards analysis as a word match and looking at your department short list of learning targets and simply finding them within the larger standards list. This is a misstep, as it will not lead toward additional growth and doesn't foster a team mentality that is focused on problem solving.

Another way to think of content standards is to frame them as *desired student outcomes*—the concepts that you want your students to have mastered at various checkpoints throughout the year. Conversely, process standards will help your students get to this point—your process standards are essentially the vehicle that helps students get to these learner outcomes. These should be comprised of a variety of focused tasks that are supported via carefully scaffolded literacy strategies. Instruction on process standards should use carefully sequenced literacy strategies that focus on specific tasks aimed at mastering your power standards. For example, teaching students to read for details that help them fully understand a problem statement is essential for the student's clear understanding.

Identification of Students' Literacy Skills

To dive into disciplinary literacy, your team needs insight into your students' literacy skill strengths and weaknesses. Literacy-based inventories or assessments, like the grade-level exams states often issue or even college-preparatory exams like the ACT or a reading passage from the SAT, will help you identify where students are now in terms of literacy skills so that you can later determine the scaffolds to implement when students must navigate challenging text tasks. While mathematics teachers may be unaccustomed to examining results from reading assessments, they will provide valuable insights into the range of reading abilities within a classroom. Knowing more about your mathematics students' reading abilities will help course teams gain a sense of where there may be gaps with various reading tasks in the mathematics classroom.

Many schools use the response to intervention (RTI; Buffum, Mattos, & Malone, 2018) or multitiered system of supports (MTSS; National Center on Intensive Intervention, n.d.) frameworks already have literacy benchmark assessments in place, although the data are not always shared among all departments in the school. These assessments typically hold a wealth of information pertaining to student strengths and weaknesses for teachers of all content areas. If your school has a reading specialist on staff, they would be a great colleague to approach to learn what data your school has already collected and what other resources might be available.

While working with a reading specialist, your team might want to design your own literacy-based inventories, employing the content area's authentic texts and assessing the literacy skills most pertinent to the specific field of study. Studying this initial student data will be a key component to your team time and will help you identify which strategies will scaffold student success and help you further develop and utilize a variety of assessment forms, including formative and summative tools. Word-based problems are the most prevalent reading task used in most mathematics classrooms, though different mathematics subjects may require reading skills that are unique to each discipline. For the format or formats that are most prevalent in the team's course, consider implementing a team-created baseline assessment early in the year that will provide the team with valuable information about students' mathematics reading skills.

Confidentiality and professionalism are necessary for a productive collaborative model. Getting to know your students better means looking at and sharing your student data with your team. When examining and collaborating around assessment data, it's important for teams to maintain a judgment-free zone, creating a safe space for team members to confront any issues the data reveal. During team meetings, it is critical that no one makes sweeping statements or generalizations regarding teaching practices based on raw data. When viewing group data in a team discussion on progress, remove names or class cohort information. Start the data discussion by noting this information may reveal things the team has already considered. For example, rarely does a nationally normed assessment show that a star student is one of the most struggling readers. Begin the data conversation with the idea that the data will often confirm what you already know about your students, but it may also shed more light on why students are struggling.

As your colleagues become more comfortable with data reviews, any fears about sharing data among team members will gradually drop away, and the team will become more open to collaborating on student learning, sharing, and growing together professionally. Examining student data can sometimes feel like finger-pointing if teachers on the team don't feel as though trust has been established. We encourage teams to create and abide by norms to ensure that team members can trust one another inside and outside of the meeting space. Maintaining the mindset that we are working with *our students*, rather than *my students*, can help ensure that team members will seek to support one another. If you are a team leader, start with yourself; don't be afraid to show your own student data. Students have entire academic histories prior to meeting their current roster of teachers.

There is no "what that one month in Mr. Williams's class" that dictates all aspects of a nationally normed score.

Although it is crucial that teams view data in a supportive manner, make sure that your team's analysis doesn't look like a list of excuses. It is very easy to fall into the pit of "things we can't control" when looking at less-than-ideal numbers. Instead, use undesirable data outcomes to examine what your team *can* change; the discussion should focus on how you move students toward a goal, target, or outcome—and ultimately toward graduation, higher education or training, and a career.

Identification of Content-Literacy Connections

When working with various content teams, we have often heard teachers comment, "There really isn't any reading in my subject area." This view is based on a traditional and outdated view of the concept of literacy. When we refer to *literacy*, we mean the act of engaging, knowing, and ultimately being able to navigate new understandings of known and unknown nuances associated with defined content. Yes, this sounds complex! But really, teaching literacy is as much about breaking down an idea, only to build it back up again by scaffolding and modeling a process—just like teaching a child how to tie a shoe.

All this said, words are literally everywhere, and many literacy texts are multimodal. Formulating an inference from a reading is similar to formulating an inference from what someone might say from a film or dramatic performance. In comparison, literacy strategies we use to understand or to infer are often similar no matter what the modality or concept might be. The strategies we provide in chapters 3 through 6 apply across content areas and can serve as a vehicle to transport comprehension.

After your team establishes a solid common baseline knowledge of the concept of literacy, focus on your power standards to identify which literature tasks are the most innate to your area of study. Here is where you can identify your process standards and consequently outline the scaffolds that will support students' literacy skill development. Additionally, you will need to take a close look at your texts or problems (solo or as a team for common texts) to determine if they are appropriate for your students and curriculum tasks. Gather all texts and assess your collection as a whole, using the tool in figure 1.2 as a guide.

1. Does the content match your defined content power standards? If not, what is missing?
2. Are these texts at an appropriate text-complexity level for your readers?
3. Are your text tools varied to address the different types of content or texts associated with your discipline?
4. Once you have gathered your collection of materials, what scaffolds will your students need to comprehend the content successfully?
5. What scaffolds will your students need in order to transfer knowledge to new contexts and applications?

Figure 1.2: Review tool to discern whether text tasks match complex disciplinary literacy demands.

Visit ***go.SolutionTree.com/literacy*** *for a free reproducible version of this figure.*

Wrapping Up

If a fully supported and committed team sounds like a far-fetched dream—that's OK. All you need is one colleague to join you in your efforts to formulate collaboration that leads to positive change! While your work may take a narrower lens, know that any form of action research is worth the effort; be sure to share your journey and findings with your colleagues. Collaboration isn't always as formal as a designated team time—it often starts with an individual teacher simply sharing what you are working on and trying to accomplish with students. By conducting a thorough analysis of your curriculum standards, texts, and students' skills, your team will be well on its way to creating a common understanding of where your students are now and be able to better determine how to move them toward disciplinary literacy.

Effective teams work together diligently and value all contributions in their quest to help students succeed. Building a productive team is essential to working collaboratively toward teaching literacy skills embedded within your content curriculum rather than only the study of literature. To do this, collectively unpack the mathematics reading, writing, and language standards, set appropriate goals for students, and develop tasks that foster student growth.

Collaborative Considerations *for* Teams

- Who are colleagues you can approach to begin fostering collaborative teamwork?
- What mathematics literacy skills lend themselves to prolonged examination for a team? Are there recursive skills that will show up throughout the year? (Because there are so many, you will need to prioritize.)
- If you already have a mathematics-specific or literacy-based team, how is it organized? Does your PLC culture support dialogue and all ideas that might lead to improved student outcomes?
- How can you use the resources in this chapter to develop your PLC culture?

CHAPTER 2

Foundational Literacy Triage

Throughout this book, we focus on how to establish a culture of literacy in the mathematics classroom as well as throughout the school. All educators committed to learning must recognize the value of literacy, and all need to celebrate the results when an entire school establishes goals and creates an environment for students to feel safe when attempting and learning literacy strategies. However, this process takes time, and we often have teachers come to us who need immediate assistance. They have specific students, classes, projects, or assignments they are struggling with, and they need a strategy or way to help the students *now*.

Always keep in mind the array of learners in your classroom. When instructing students in a mathematics classroom, there will be a range of student learning proficiencies. Further, students' reading levels may be even more diverse. Among these reading levels, you will have students with special education needs (emotional, learning, or medical), students with 504 plans, and students who are only just learning English. All of these student groups are capable learners, but they all have different needs as unique as the learners themselves. As mathematics teachers, your team must meet the needs and develop the potential of *all* the students in your classes.

We have seen that teachers often begin to notice concerns related to literacy early in the school year and look for differentiated instructional ideas to meet student needs with the necessary supports. This chapter helps teams identify ways to triage literacy within mathematics classrooms immediately and work with the variety of student needs and abilities found in your classrooms. In this chapter, we cover the basics of RTI practices, differentiated instruction, assessment of text complexity, and several fix-up strategies to assist mathematics students when reading a text (Tovani, 2000). We close by examining factors to consider when supporting striving learners and when enriching the learning of students who show proficiency.

Understanding Response to Intervention

Response to Intervention (RTI), a multitiered system of support (MTSS) for academics, is a three-tiered systematic process for ensuring the time and support students need in order to learn at high levels. Austin Buffum, Mike Mattos, and Janet Malone (2018) explain, "Tier 1 represents core instruction, Tier 2 represents supplemental interventions, and Tier 3 represents intensive student supports" (p. 2). Figure 2.1 illustrates the inverse pyramid Buffum and colleagues (2018) present for RTI practices.

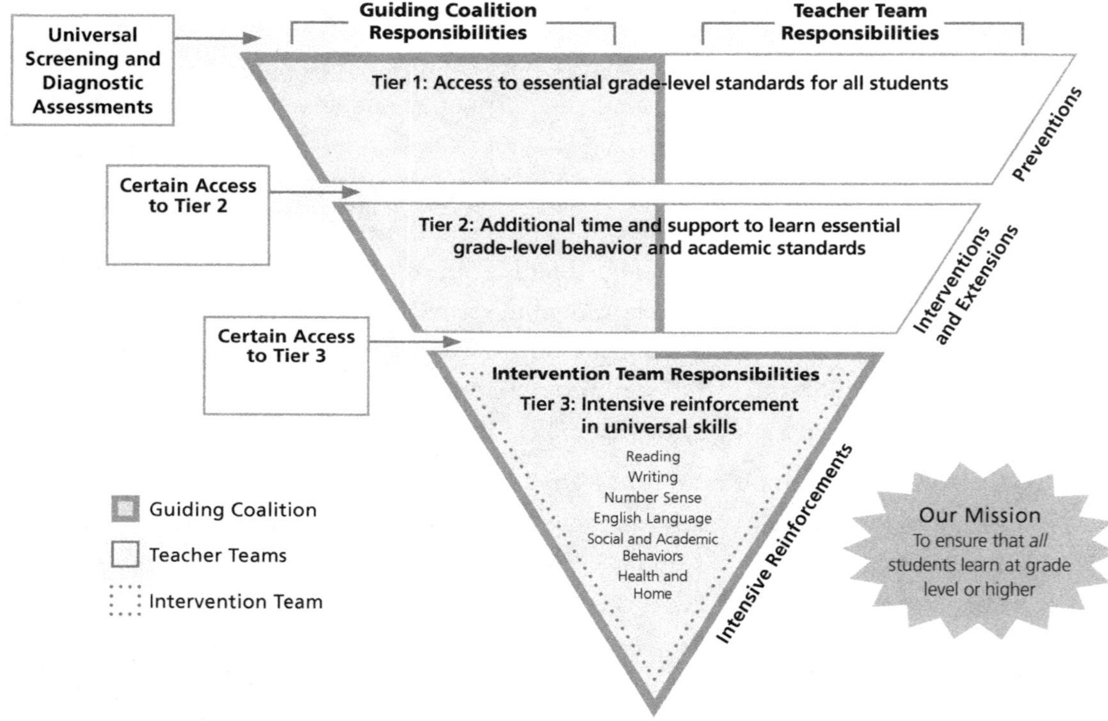

Source: Buffum et al., 2018, p. 18.

Figure 2.1: The RTI at Work pyramid.

Those familiar with more conventional RTI content may not be used to the inverted pyramid of RTI at Work. Of this inversion, Buffum et al. (2018) write:

> The traditional pyramid seems to focus a school's intervention system toward one point: special education. Subsequently, schools then view each tier as a required step that they must try to document prior to placing students into traditional special education services. Tragically, this approach tends to become a self-fulfilling prophecy because the organization starts interventions

> with protocols designed to screen and document students for this potential outcome.
>
> To challenge this detrimental view of the traditional pyramid, we intentionally inverted the RTI at Work pyramid, visually focusing a school's interventions on a single point—the individual student. (pp. 18–19)

Because a student's literacy skills are often linked to their learning in other academic areas, thinking about the role of literacy in RTI is an important topic to cover when students need support. We believe that teaching literacy is the responsibility of all teachers, and we think working with literacy interventions is a Tier 1 commitment—that is, all students should have access to grade- and course-level instruction related to literacy. As we noted earlier, teams often need more intensive literacy-based strategies to support students who might struggle or challenge students who excel—this is where Tier 2 or Tier 3 interventions, remediation, or extensions might be necessary. Identifying students who will benefit from more intensive support in areas of literacy is an important consideration early in the school year—especially in a mathematics classroom, where the reading and writing can look different than what students encounter in other disciplines and thus can challenge students.

Gathering data about students' reading and writing skills can be helpful in learning more about the range of abilities in the mathematics classroom. Traditionally, literacy and reading data are examined by ELA teachers, while mathematics teams may rely on mathematics-based screening data. Data found on nationally normed tests, such as MAP (www.nwea.org) and STAR (www.renaissance.com), can provide mathematics teachers with early insights into students' literacy skills. We believe that both of these types of assessments—literacy-specific and mathematics-based—can provide invaluable information about individual students in the mathematics classroom. Likewise, teams should collaborate around creating common formative assessments that focus on measuring literacy-based skills important to the mathematics curriculum they are teaching. Throughout this chapter, we outline a number of literacy-based concerns teams should give their attention to as students are starting the school year, and we offer suggestions for responding if and when you and your team notice other concerning patterns in student learning.

Differentiating Instruction

A good place to start with meeting students' immediate, varied needs is to consider how you can differentiate instruction. *Differentiated instruction* is instruction that helps students with diverse academic needs and learning styles master the same challenging academic content (Kise, 2021). It also provides students with interrelated activities that are based on student needs for the purpose of ensuring that all students come to a similar grasp of a skill or idea (Kise, 2021). Differentiating does not mean providing separate, unrelated activities for each student. Instead, differentiation supports students with learning differences, helps students retain content and skills essential for learning, reduces time students require to learn information, decreases the need for skill remediation, and allows students to demonstrate their learning in a variety of ways.

There are four planning steps you can use when differentiating classroom work.

1. **Determine the academic concept or skills:** The first step in differentiating instruction is knowing and understanding the specific mathematical skills you want to teach. This means knowing what learning targets (process standards) you want students to accomplish. What is the end result you want for your students to know and understand within this mathematical concept?
2. **Gauge students' background knowledge:** You need to have a basic understanding of what the students currently know about the mathematics topic or content. At times, the students may have no background knowledge on a topic. Consider issuing a benchmark assessment or survey that will help to determine students' readiness for new content.
3. **Select suitable instructional methods and materials:** Once teacher teams have an understanding of what the students already know, instructional methods need to bridge what students currently know about the topic and what you want them to learn during the lessons. Knowing what knowledge gaps students have and what information they lack as well as where they are already proficient will help you plan to use specific methods and materials that are appropriate for instruction.
4. **Design ways to assess skill mastery:** After delivering instruction in multiple ways, the last step is to assess students' accumulation of new skills and knowledge for mastery. These formative assessments—assessments that move the learning forward—can take many different forms and be as simple as an exit slip or a quick response to a learning prompt. We cover assessing mathematics students in chapter 7 (page 107).

When working with these four steps of differentiation, the process overlaps. Students all bring different knowledge to the content, work at different paces, and understand the material at different times. The overlap makes it possible to tailor the process to fit each student's needs at a particular time. In the following sections, we examine each of these steps in greater detail.

Determine Academic Content or Skills

As described earlier, PLCs focus teams on four critical questions. The first thing a team should ask is, What do we want all students to know, understand, and be able to do? (DuFour et al., 2016). As a mathematics teacher approaching these questions, determining the academic content or skills (or both) is the first step toward focusing decisions about curriculum, instruction, and assessment choices. We find that aligning literacy-based strategies to support these choices will help teams support students' learning of mathematics. As you and your team work through the strategies in this book, consider how certain strategies are better suited for the academic content and the skills you and your team are focused on teaching students. For instance, if you are working on students' ability to create equations using multiple variables after reading a word problem, your team should work with strategies that help students closely read the text to bridge the reading skills with the mathematics skills.

Gauge Students' Background Knowledge

One challenge in teaching mathematics is that it's common for students to have limited background knowledge of mathematics, making the teaching of new skills particularly difficult. Content standards in mathematics are often linear in nature, so if a student has yet to master a standard from a previous mathematics course, it can hinder their path to success in their current course. When students' background knowledge is limited, they struggle to make immediate connections or draw quick associations to what might be familiar to them, making it a challenge for them to add to their existing understanding or depth of knowledge. Mathematics is often abstract and unfamiliar to students, so teachers have an obligation to meet students where they are when introducing new mathematical concepts. Many students carry a deficit mindset into the mathematics classroom, believing "I'm just not good at math." This could be for a number of reasons: a lack of procedural knowledge, skill deficits, or emerging literacy skills, to name a few.

When students have limited background knowledge about mathematics topics, teachers and teams must recognize that they will need to approach prereading strategies in ways that can draw connections to students' interests and their limited understanding of a topic in mathematics that might be new to them. Teachers and teams can access students' background knowledge using the prereading strategies in chapter 4 (page 63). While the use of these strategies presents an extra step for classroom instruction, they present a creative option to merge literacy and disciplinary learning.

Select Suitable Instructional Methods and Materials

Selecting suitable instructional methods and materials demands that teachers and teams ask a few different questions. For example, when considering literacy skills, *selecting suitable materials* means aligning reading and writing tasks to the appropriate ability level of your students during their grades 6–12 experiences in mathematics. Pay close attention to the reading level of the materials you select for students; they should be at or slightly above grade level. The reading level of the material can be determined by completing a readability measure. This will assist in identifying at what level the language is written and assist in student accessibility of the material. Students who require more support might need supplemental reading (problems) at an easier reading level that will scaffold them up to the grade- or course-level material. Students who are more proficient readers might benefit from reading (problems) that extend their understanding of the mathematics content. Remember, that reading looks different in different content areas. Reading in mathematics is interpreting, analyzing, and solving a problem. The following example executes the scaffolding of a mathematics problem for students to access, understand, and solve.

> Figure A and figure B are both regular polygons. The sum of the perimeter of figure A and the perimeter of figure B is 63 inches. The equation $3x + 6y = 63$ represents this situation, where x is the number of sides of figure A and y is the number of sides of figure B. Which statement is the best interpretation of 6 in this context?
>
> a. Each side of figure B has a length of 6 inches.
>
> b. The number of sides of figure B is 6.
>
> c. Each side of figure A has a length of 6 inches.
>
> d. The number of sides of figure A is 6.

In the preceding problem, the instructor can read the question to the students and then begin working thought the reading and solving of the problem. The process is as follows:

- Read the problem.
- Underline important information or vocabulary words in the problem (such as *polygon*, *perimeter*, *x*, *y*, *interpretation*, and so on).

- Define unknown vocabulary.
- Identify what the problem is asking.
- Set up equations.
- Solve for the answer.
- Check your answer.

Creating an outlined reading process makes a more abstract concept more concrete and approachable. Creating various lists depending on the mathematical concepts is an adjustment the instructor will need to identify.

One benefit of working on a team is that members collaborate to locate a variety of possible readings that assist in helping to interpret, analyze, and solve a problem to help all students advance in their abilities. Likewise, drawing on the expertise of a team can help inform the differentiated instructional methods each team member uses to help all students develop to their fullest potential in mathematics.

In considering instructional choices, focus your team on discussions that can help support students in ways that advance their ability to confidently approach simple texts and simple questions, and complex texts and complex questions, with greater confidence. In our work with mathematics teachers, we find this approach to differentiated instruction helps all students, illustrating that there is no one-size-fits-all instructional strategy. When creating original materials—specifically word problems—we find that students thrive when teachers and teams write about topics that students are interested in. Not only does this help students to develop mathematical skills, but it builds a classroom community where students feel seen. Examples of problems using topics of interest can be relating the problem to current pop culture, sports, activities, and current-event content, which will engage students in the text more readily.

Design Ways to Assess Skill Mastery

Assessing skill mastery addresses the second critical question of a PLC: How will we know if they have learned it? (DuFour et al., 2024). In our work around literacy, we encourage teachers and teacher teams to focus on assessing skills in formative ways, often and over time. Although we specifically cover assessment strategies in detail in chapter 7 (page 107), many of the strategies we present throughout the book have applications in formative assessment. When teachers work with formative assessment practices, they can respond more immediately to the learning needs of students and support students quicker and more effectively.

Remember that through your commitment to teaching, assessments are meant to encourage collaborative conversations about the literacy of students in the mathematics classroom. This collaboration helps to focus discussions on how team members can support all students and work to address the third and fourth critical questions of a PLC: What do we do if students have not learned it yet? How do we extend the learning for students who did learn? (DuFour et al., 2024). Well-designed assessments engage teams in developing and extending the potential of all students. Well-designed mathematical assessments inform what the students know both through the reading, interpreting, and analyzing lens. In addition, the students then need to solve the mathematical problem.

Consider your most recent assignment involving reading (interpreting, analyzing, and solving of a problem). How can you differentiate the reading for students at multiple reading levels within your classroom setting?

Assessing Text Complexity

An important factor to consider when seeking immediate assistance with instructing your students is making sure the reading level of the text is appropriate. This information may be available to you at your fingertips. Many times, the publishers of a book provide a Lexile measure or reading-level score, and if a resource doesn't, you can usually determine its level using search tools at the Lexile website (https://hub.lexile.com). Identifying the level of the text helps teachers understand if the student is struggling with interpreting the problem or specifically the mathematical calculation of the problem. Equally important is to know the Lexile level of your students. Having this knowledge allows you to appropriately select materials that will challenge students but not frustrate them—materials that are at or slightly above their Lexile level. You can determine individual student Lexile levels through standardized testing measures. (Visit https://bit.ly/2ZmS5EF to learn more about this.)

With knowledge of each resource's Lexile level and the Lexile level for each of your students, the challenge for teachers and teams is to provide a range of texts on a topic that fits the differentiated needs of your students so that no text is too complex (students do not learn the material needed to bridge the gaps in their learning) or too basic (students do not advance their knowledge during their learning time). Using texts at the students' instructional level helps to differentiate if individual students are being successful with the mathematical concepts due to their interpretation of a problem or calculations. These can be used directly with instructional material or with supplemental material. If the Lexile level is above a student's abilities, teachers can support the student to clarify mathematical text by defining words in the context, either in parenthesis or by providing synonyms so students can make the connection.

Using Fix-Up Strategies

Fix-up strategies—strategies to help students get unstuck during independent reading when they find a text confusing—are another way to assist mathematics students when reading a text (Tovani, 2000). They provide a way to not only model but also teach good literacy habits and strategies to students. Consider how you could apply some of the following strategies to a word-based mathematics problem. Cris Tovani (2000) writes that modeling in this way can support reluctant readers by providing more explicit instruction and reminders to use fix-up strategies, which include the following, phrased in student-friendly language.

- **Make a connection:** Find connections with what you are reading to other texts, experiences you have encountered, or the world around you. Making a mathematic problem relevant to updated topics will help the students make a connection to what they are reading.
- **Make a prediction:** Predict what will come next in a text or make a prediction prior to solving a word problem.
- **Stop and think:** When something doesn't make sense, stop and think about what you've just read or the step you've just taken in a problem.
- **Ask yourself a question:** When something doesn't make sense, ask yourself a question about the text. Try to answer the question using what you know or new information you've found in the text. Self-questioning assists students while they are working on a problem to understand steps.
- **Reflect in writing on what you've read:** Reflecting on your own thoughts about what you read can help to solidify connections to a reading and identify what might be new. It can also give you the opportunity to formulate an opinion. Reflect on what you might agree

with or disagree with, on what stands out in the reading and why, or on what makes you think differently about the topic and why.

- **Visualize:** Stop and create mental images of the concepts described in the text or the given problem.
- **Use print conventions:** Examine headings and emphasized words to identify important ideas in the text. Identify what words are important in solving the problem.
- **Retell what you've read:** To help build the ability to summarize and synthesize, after you read something, stop and retell someone (or yourself) what you read in your own words. Try to capture the key point and key details in a way that makes clear sense to you, the reader.
- **Reread:** If something doesn't make sense, reread for clarity. Start at the beginning of the problem and work through all the steps again.
- **Notice patterns in text:** Consider whether reading is establishing a pattern. For instance, does the reading argue both sides of a debatable topic, and are there patterns to how the reading is presenting those viewpoints? Do patterns in the reading establish cause-and-effect relationships? Why or how does the pattern help guide the development of the reading?
- **Slow down or speed up:** Adjust your reading to slow down or speed up depending on your level of understanding.

Figure 2.1 provides a sample bookmark students can use for reference.

Fix-up strategies are always a good reminder of how to read a text or mathematical problem. You can create bookmarks for all of your students, listing these strategies as a quick reference to use when reading. A tangible bookmark in their hands when reading will help remind them to use literary strategies consistently. When they do so, these strategies will become habits students can apply whenever they struggle with reading comprehension or how to solve a problem.

When using the fix-up strategies, we recommend several different ways to make sure students are using them effectively.

- Help students realize that *during-reading strategies*, like those we present in chapter 5 (page 81), are important *skills* to learn. Developing during-reading skills can help students with their comprehension skills and critical-thinking skills. When first teaching these strategies, help students to understand the value the strategy has for their learning.

- Model how to use the skill. Try not to assume that students will simply use the strategy. Take the time to model *how* the strategy is used. For instance, to model how to read and ask questions or how to read and retell, a teacher might complete an oral read-aloud of their thought process while directly instructing a problem. Often, students can practice these during-reading strategies with collaborative partners in class. With all the mathematics-specific strategies we present in the forthcoming chapters, modeling each of the habits is also good practice, and it provides students with examples of what they can do and what you expect of them. Be sure to celebrate when you observe a student using any of the strategies.
- Encourage focusing students on using one or two during-reading strategies at a time. Consider the reading you are giving students, then align one or two strategies that will be most effective for that particular reading. Ask students to practice that strategy.

It's important to remember that you and your team can adapt all of these habits or strategies to meet the specific needs of your students in the immediate classroom setting. For instance, a teacher can adapt and use the visualizing strategy for use in small groups or with partners, where a group can create a drawing or image to demonstrate their understanding of a mathematical concept.

FIX-UP STRATEGY BOOKMARK

★ Make a connection.

★ Make a prediction.

★ Stop and think about what you are reading.

★ Ask yourself a question.

★ Reflect in writing on what you've read.

★ Visualize: Stop and create mental images of the concepts described in the text.

★ Use print conventions: Examine directions and emphasized words to identify important ideas in the text.

★ Retell what you've read.

★ Reread: If something doesn't make sense, reread for clarity.

★ Notice patterns in text.

★ Slow down or speed up: Adjust your reading to slow down or speed up depending on your level of understanding.

Figure 2.1: Fix-up strategies bookmark.

*Visit **go.SolutionTree.com/literacy** for a free reproducible version of this figure.*

thinking
BREAK

What additional questions or thoughts, at this point, should you consider that can specifically assist in the mathematics classroom? When collaborating with your team, what additional resources will you need to help move literacy in the mathematical classroom forward?

Considerations When Students Struggle

Some students will likely continue to struggle and need repetition with these strategies even after you implement differentiation, find or adjust appropriate texts to support their reading levels, and add additional steps to their direct instruction to define specific terms and share fix-up strategies. As with implementing any new strategies or processes, these struggles may persist as students begin to learn how to read and process a mathematics-based text (interpreting, analyzing, and solving a problem) in ways that help to develop a more attentive level of comprehension and the ability to think *while* reading. This takes practice and commitment, which is why we think all teachers should work with students on these strategies for learning. Some additional questions to consider when you encounter students who are having a hard time accessing the text or using basic fix-up strategies include the following.

- **How might you help students recognize what level of text they are reading or able to access?** Many secondary mathematics teams utilize a textbook, so it's imperative to evaluate the reading level of the text. If the student struggles with too many of the vocabulary words in the text, the reading might be too difficult for them, and they might need support. Mathematics texts are often filled with new vocabulary words when introducing new mathematics concepts, so working on vocabulary development in mathematics readings is often an important commitment to student learning.
- **How might you provide immediate help to students who have trouble reading or understanding specific mathematics vocabulary?** There are many ways to improve vocabulary development. The most important way is by having the students experience the word in differing ways and contexts. For example, a teacher can add synonyms to individual words using parenthesis. This provides additional words for the student to use when reading the content. Likewise, students need to *use* the new words

they are learning. Consider building *word walls* around your classroom—charts of paper that list important vocabulary words students should be learning to help support their comprehension and understanding of mathematics and mathematical concepts. Refer to the word walls and ask students to use the words regularly during class time and within their group work. Remind them that in order to think like a mathematician, they need to talk like mathematicians.

- **How might you help students stay focused during reading and use the fix-up strategy bookmark you provide?** Have students use the bookmark as a reminder or reference for what students should be thinking about while they read. When establishing a purpose for reading, ask students to refer to the bookmark as a way to keep their focus on that purpose or on the key questions they should be considering.

Although these strategies are helpful, some students will continue to struggle with the demands of literacy, especially as mathematical texts, concepts, and principles increase in difficulty. When working with students who might continue to struggle, it is important to intervene and anticipate difficulties they might confront.

As an individual teacher and as a team, work to prepare supports that will help every reader to succeed with comprehending and understanding the literacy tasks expected of them. In setting expectations for reading that are directed and purposeful, teachers can help students achieve at a rate that is individualized through appropriate challenges. A mathematics reader who is still developing might need to focus on more simple questions during a first reading and break things down step by step of what is being asked, and then focus on a more complex question or combine the questions during a second reading. Or a developing reader might need to be directed on three important points to identify so they can then find them in the mathematics text. In working with varying reading levels, team collaboration can help with generating new and fresh approaches to the assigned mathematical literacy tasks.

Considerations When Students Are Proficient

Just as it is important to support the developing reader in your mathematics classrooms, as teachers, collaborative teams also want to develop the potential and critical-thinking skills of students who demonstrate proficiency and mastery. When working on literacy-based tasks, consider working with your team to create more complex questions and tasks that will challenge students to extend their critical thinking capacity. Your team might have other readings that advance students' disciplinary knowledge, but you might also ask the students to process their thinking about a text in more complex ways. For instance, ask students to respond to higher-order-thinking questions that are focused on building connections or associations between concepts in mathematics, or ask students to grapple with problems in different forms. For example, the teacher may not provide as many of the givens in a problem or outline things as directly. This provides an opportunity for students to use their metacognition when approaching and solving the problem.

As a team, remember that your collaborative work should address the developing potential of all students. Often, by focusing on more advanced ways of thinking, teachers and teams will uncover creative and interesting insights into revising their curriculums to be more engaging, too.

Wrapping Up

By establishing a triage approach to differentiating instruction and finding texts at students' varied reading levels, your team establishes a powerful way to open up new opportunities for the reading process. If you match this with fix-up strategies and share a bookmark listing these strategies, you can equip students for a more successful reading process and help them gain the confidence to continue reading their mathematical text and problems. Modeling the fix-up strategies is key to showing students how to use them. Continuing to provide opportunities to practice and repeat strategies will help students learn the strategies as ongoing habits. In this way, the strategies become more automatic to their own maturing reading process.

Collaborative Considerations *for* Teams

- What fix-up strategies seem most appropriate for a mathematical text?
- How might you go about adapting the texts and problems you use for different reading and ability levels?
- What are ways your team can use or adapt mathematical texts to challenge students who have already demonstrated proficiency?

CHAPTER 3

Thinking Like a Mathematician

Although disciplinary literacy is a more visible and frequent professional development topic in the age of Common Core State Standards (NGA & CCSSO, 2010), the connection between literacy and mathematics is not immediately obvious for most mathematics teachers. When we began trying to build a mathematics team focused on literacy here at our school, many in the mathematics department were hesitant and skeptical. For many other disciplines, text readings are a regular part of the curriculum; however, in mathematics courses, the textbook does not present as "normal" reading compared to other subjects. Students need to approach mathematics texts differently, keeping the focus on interpreting and analyzing concepts and principles in order to solve mathematical texts.

This chapter explores the notion that mathematics literacy requires teachers and students to begin thinking like mathematicians. After discussing the rationale and the need for collaboration when combining literacy and mathematical expertise, we highlight the literacy skills embedded within the CCSS for Mathematics (NGA & CCSSO, 2010) and how working on thinking like a mathematician addresses some of those skills. The chapter concludes with considerations for addressing students who continue to struggle and students who reach proficiency and are ready to take their learning further.

The Importance of Thinking Like a Mathematician

The truth is that mathematics literacy is different from literacy in other disciplines. To address the understandable skepticism when we were building our mathematics collaborative team and creating a focus on literacy schoolwide, we started by demonstrating the connection between mathematics and literacy to make it more obvious and to show how literacy strategies could benefit students. Literacy expert Doug Buehl (2017) writes, "Students need to be mentored to think mathematically not only when occupied with solving mathematics problems but also when reading about mathematical concepts and relationships" (p. 106). When teachers fail to mentor students to think in this way, students retain less learned material over time (Siegel & Fonzi, 1995). Thinking like a mathematician means fluently breaking down mathematics problems into usable, logical chunks. So when students are taught to read mathematical texts closely and to think about the specialized symbols and vocabulary like the experts, they develop confidence and the ability to digest these materials. For example, mathematics uses many symbols. Code-switching from words to symbols and back to words helps students to read more fluently within the content context. Once they are able to demonstrate this skill, their confidence builds within the content and discipline.

Collaboration to Combine Literacy and Mathematics Expertise

A collaborative culture is one of the big ideas of a PLC. Since "accomplishing the goal of disciplinary literacy instruction requires that teachers be well versed in the content, discourse patterns, literate practices, and habits of mind within specific disciplines" (Fang & Coatoam, 2013, p. 7), it makes sense to bring together literacy experts and mathematics experts to learn from each other and to create innovative ways to approach student learning. After all, while content teachers often receive a general literacy background during their training, they are usually not equipped to teach literacy in depth. Similarly, most literacy specialists are not required to have specific content area expertise, so it makes sense to collaborate in teacher teams to best serve students. As literacy experts, we find that many teachers from all subject areas have some knowledge of basic and intermediary literacy skills, like the ideas we discussed in chapter 1 (page 19); however, we've also found that many teachers need support in how to adapt and develop ways to make use of literacy strategies that are specifically tailored to mathematics and the complexity level of the texts. When mathematics teacher teams have the opportunity to collaborate on high-impact strategies, they can help to ensure that all students have an entry point into a variety of mathematical texts.

To ensure quality collaboration and responsibility, literacy and mathematics experts need to enter into an equal partnership. Unfortunately, schools often prioritize content over literacy. So in order to be effective, schools must also stress the importance of literacy instruction so literacy experts do not just feel like classroom aides (Fang & Coatoam, 2013). Lucky for us, our school has prioritized literacy, so we were able to enter into our mathematics team on somewhat equal footing as educators.

During initial meetings gathering support for a mathematics literacy team in our PLC, we listened to the mathematics teachers' concerns about student struggles. According to the teachers we met with, issues that are important to them include the following.

- **Students struggling to use what they learned in the past to inform their ability to problem solve in the present:** Mentoring students to think like mathematics experts may help them more readily recall past learning and use it to solve new problems. We explore this issue in more detail later in this chapter.
- **Students struggling to articulate their problem-solving process:** Mentoring students to work through mathematical texts will give them a language to better explicate their ideas. We suggest specific strategies to encourage students to use these articulation skills.
- **Students struggling to effectively articulate solutions to problems either in one sentence or in a proof:** As literacy experts, we know that students need to write more across all disciplines to improve. Mathematics teachers should get students to write more often. This can be in the form of journal writing or specific to a mathematics concept or problem. Having students write their processes down helps them to better comprehend the material.
- **Students struggling with vocabulary that has specific mathematical meaning in addition to their colloquial meaning:** We provide some vocabulary strategies and graphic organizers to help students think like mathematicians (see chapters 4, page 63, and 5, page 81, for more specific strategies).

Soon after our initial meeting, a small group of mathematics teachers agreed to partner with us to work to address issues of literacy within the mathematics discipline. These teachers allowed us to work with them on a consistent basis in a coaching model to implement the previously mentioned strategies. The collaboration of a mathematics professional and a literacy specialist together can be powerful and rewarding.

Consider the last text students read in your classroom, whether a section of the textbook, a word problem, a geometry proof, or another text. What struggles did students encounter while reading?

- As a mathematics expert, what are the steps you might have gone through when reading the material? Would any of those steps aid students in their reading?
- How would one of your mathematics colleagues read the material? Would any of their steps aid the students in reading?
- As a mathematics team, what steps you would want students to use to successfully navigate dense mathematics texts?

How Mathematicians Think

The answer to this question might seem obvious, but when we ask mathematics teachers to explain to us how they would think through a mathematics text, it often results in an a-ha moment. The mathematics teachers articulated many more steps or thought processes than they realized. Having them write down each step is helpful in the realization of the process. Michael K. Weiss and Deborah Moore-Russo (2012) suggest that most teachers base their teaching styles on their many years of experience in the classroom. Unfortunately, for most mathematics teachers, that means thinking of mathematics as a rigid system of manipulating symbols instead of developing deep, literate thinking and reasoning mathematical skills. But now that we want to teach students to think like mathematicians, it is imperative that we dig into and expose the real thinking and reasoning that goes into problem solving.

If teachers want students to think like mathematicians, they need to show them how the insiders and experts think. To know what to teach students, it is helpful for teacher teams to collaboratively record in detail the steps they use to think through mathematical problems and texts. (See figure 3.1.) Team members can compare their steps and processes in order to create a hybrid of logical thinking to use to mentor students. When teachers model processes for students, students are better equipped to think in a logical manner about the subject at hand. Buehl (2017) refers to this type of modeling and mentorship as teaching students to think like disciplinary "insiders."

Read the following middle school algebra problem and use the following space to write each step you take as you read and solve the problem. While we have included space for ten steps, there is no exact number as each mathematician may think slightly differently.

John likes coffee a lot. The coffee at Great Grounds costs $5.25 a cup. If John buys coffee every day for the month of September, how much will he spend?

a. Coffee costs $5.25 a cup.
b. The month of September has 30 days.
c. Multiply 5.25 × 30.
d. The answer is $157.50.

1. ______________________ 6. ______________________
2. ______________________ 7. ______________________
3. ______________________ 8. ______________________
4. ______________________ 9. ______________________
5. ______________________ 10. ______________________

Figure 3.1: Steps to think through a mathematics problem.

The problem in figure 3.1 is a simple algebra problem; however, the process is important to solve the problem and more complex word problems correctly.

CCSS for Mathematics Connections

As we discussed previously, six of the eight CCSS Mathematical Practices (NGA & CCSSO, 2010) have a significant literacy component. We highlight these anchor standards rather than specific grade-level standards because this book is for grades 6–12 mathematics teachers, and the complexity of each standard varies in different grade bands. The Mathematical Practices highlight the literacy connections that appear throughout grade levels.

- **CCSS.MATH.PRACTICE.MP1: "Make sense of problems and persevere in solving them"** (p. 6). This standard requires an inner dialogue where mathematics-literate students explain to themselves the meaning of a problem, analyze the parts of a problem, explain components of and relationships within problems, and metacognitively monitor their progress toward a solution.

- **CCSS.MATH.PRACTICE.MP2: "Reason abstractly and quantitatively"** (p. 6). This standard requires students to conceptualize relationships between mathematics symbols and meaning.
- **CCSS.MATH.PRACTICE.MP3: "Construct viable arguments and critique the reasoning of others"** (p. 6). Not only are students required to build, analyze, and compare viable arguments and solutions for problems, but they must also be able to look at the arguments and solutions of others and explain why those are or are not viable.
- **CCSS.MATH.PRACTICE.MP4: "Model with mathematics"** (p. 7). Mathematics-literate students can use mathematical concepts to solve real-life problems and models. This means students must design problems, often by writing them out with mathematical symbols and language, and then solve and use the application.
- **CCSS.MATH.PRACTICE.MP5: "Use appropriate tools strategically"** (p. 7). Students must be literate enough to read a problem deeply and decide what materials are best suited to solve it—from paper and pencil to technology—depending on the problem at hand.
- **CCSS.MATH.PRACTICE.MP6: "Attend to precision"** (p. 7). Students are required to communicate precisely. They also are asked to precisely examine claims and use precise mathematics terms and definitions.

In this chapter, it is fair to say that thinking like a mathematician involves all of the six preceding standards. Mathematicians have to make sense of problems, reason through them, compare and contrast options for solving problems, relate mathematics to real-life situations, use appropriate tools for solving them, and clearly explain their solutions. Thus, by mentoring students to think like mathematics experts, mathematics teams are teaching students these important standards. For the strategies that follow in this chapter and in chapters 4 (page 63), 5 (page 81), and 6 (page 97), we use a chart to show which strategies address which CCSS Mathematical Practices.

Strategies for Supporting Mathematics Students to Think Like Mathematicians

As literacy coaches, we often begin our coaching sessions with content teachers by asking them how they go about thinking through reading or problem solving (as in figure 3.1, page 53). When content experts, in this case mathematics teachers, identify their own thinking processes, they are better able to identify how to mentor students in thinking processes for their specific discipline. In this section, we explore two strategies: The Mathematics Process Log and text-dependent questioning. Table 3.1 shows how these strategies apply to the CCSS Mathematical Practices.

Table 3.1: Mathematics Process Log Graphic Organizer Addressing Six of the CCSS Mathematical Practices

Strategies	MP1	MP2	MP3	MP4	MP5	MP6
Mathematics Process Log	x	x	x	x	x	x
Text-dependent questioning	x	x	x	x	x	x

Mathematics Process Log

Early on in our work with algebra teachers (predominantly in ninth grade), the teachers identified word problems as a student area for growth. With the shared goal of improving students' ability to read and solve word problems, we set to work. One additional resource our students have is a literacy support course for those identified as in need of additional support. This course is designed to work on literacy across all disciplines using a unit of study from multiple disciplines across the school. Mathematics is included in this course, which allows for collaboration between ELA teachers, mathematics teachers, and literacy specialists, and for students to work on mathematics literacy through mathematical vocabulary, processes, and ideas. Since many struggling students also took our literacy support course, we brought in the literacy teachers to work with the mathematics teachers to develop a way to mentor students in thinking through word problems.

Later, when working with an honors geometry teacher, we engaged in a similar thinking process. After being exposed to Buehl's (2017) text-dependent questioning strategy from *Developing Readers in the Academic Disciplines* in one of our large-group professional development sessions, the geometry teacher realized that by walking his students through the process of talking to and asking questions of the text, he could better prepare them to think through problems like mathematicians. Together, we adapted Buehl's (2017) original graphic organizer, intended for text reading, into a thinking process graphic organizer for mathematics students. (See figure 3.2, page 56). Note that the log includes three types of actions: (1) presolving, (2) solving, and (3) postsolving.

Mathematics Process Log: Presolving, Solving, and Postsolving	
Question: John likes coffee a lot. The coffee at Great Grounds costs $5.25 a cup. If John buys coffee every day for the month of September, how much will he spend? **Extra challenge:** How much would John spend if he buys a cup per day for a year (365 days)?	
Presolving: Read the problem and re-state the problem as clearly as you can.	How much money will John spend if he buys coffee every day in September?
Presolving: Reread the problem and explain how many steps are involved in the problem.	1. Figure out an appropriate equation. 2. Find the appropriate numbers and data. 3. Put the numbers into the equation. 4. Solve the problem. 5. Write out the answer.
Presolving: Reread to identify what mathematics operations will you use.	price x days = total cost
Presolving: Reread to see if the problem has special difficulties or things you have to watch out for.	How many days are in September?
Solving: What do you do first?	**Figure out the data:** Coffee costs $5.25 per day; there are 30 days in September.
Solving: Then what?	**Put the numbers in the equation:** $5.25 \times 30 =$
Solving: Then what?	**Solve the problem:** $5.25 \times 30 =$ $157.50
Solving: Then what?	**Write an answer:** If John buys a coffee every day in September, he will spend $157.50.
Postsolving: What do you do to check your work?	**Check your answer:** I can rework the equation to see if it holds true. $157.50 divided by 30 days should equal the cost of the cup of coffee and it does!

Source: Adapted from Buehl (2017).

Figure 3.2: Sample completed Mathematics Process Log.

Visit ***go.SolutionTree.com/literacy*** *for a free reproducible version of this figure.*

The Mathematics Process Log (figure 3.2, page 56) walks students through a logical progression of thinking used by experts to solve mathematical problems. Regular use of such a process should help students get in the habit of slowing down to solve problems with prethinking instead of simply jumping into the numbers and looking to quickly solve them without any context. Students who ignore the reading of the problem and the prethinking may be prone to making careless mistakes or misunderstanding what the problem is asking of them. By training students in pre-, during-, and postsolving thinking, we set students up for successful thinking like a mathematician.

How to Use

Since the steps are easily broken down into pre-, during-, and postsolving sections, we recommend that teachers work on each of the three sections slowly and deliberately. Start by modeling the presolving steps without fully working through the during- and postsolving steps. Allow students to dig into the thinking that should occur before even attempting to solve a problem. Once students are comfortable with the prethinking needed to work on problems, walk them through the steps to solving problems. After gradually releasing them to practice presolving and solving on their own, end with a focus on the postsolving steps. There is no reason to rush through this strategy as students should come to view mathematics as an engaging thinking process that develops pre-, during-, and postsolving thinking.

Adaptations

When adapting the Mathematics Process Log for students who are struggling with the literacy aspect of the strategy (identifying what the problem is asking them or the step-by-step process), it can help to initially provide sentence starters or hints in the writing column to prompt student thinking. For example, the teacher might prompt students with "How many days are in September?" It may also be helpful to have struggling students articulate their thoughts out loud or explain their thinking with a partner. Hearing the thoughts aloud helps students better understand confusing ideas or concepts and clarify different thought processes. There are specific apps teachers can use to explore the concept of having students explain their thought processes. These include Explain Everything, Doceri, MindNode, or MindMeister. Encourage students who consistently and accurately provide evidence of proficiency with the Mathematics Process Log. (See page 129 for a blank reproducible version of this tool.) Students can use this tool to create sample problems and record their thinking to provide models for other students to use for support or studying. A variety of sample problems will help students see multiple ways of thinking through the problem-solving process.

Text-Dependent Questioning

Similar to the Mathematics Process Log, the Text-Dependent Questioning strategy and graphic organizer supports students in walking through a logical progression of thinking used by experts to solve mathematical problems (see figure 3.3). By asking questions about the problem, students attempt to discover the information they need to adequately solve it. This tool includes basic questions—What does the problem tell you?, What does the problem give you?, What connections can you make?, and How does this apply to broader mathematics?—supported by prompting questions.

Also, like the Mathematics Process Log, regular use of such a strategy helps students get in the habit of slowing down to solve problems with prethinking instead of simply jumping into the numbers and looking to quickly solve the problem without any context. By thinking deeply about the problem, students are more likely to avoid errors, and this strategy also encourages students to think about connecting problems to what they already know and to think about how the problem might apply to other situations. The end goal, again, is to encourage students to think like mathematicians.

How to Use

The best way to use this strategy is to model each section and gradually allow students to build individual proficiency. Keep in mind that it is not always necessary to solve the whole problem, especially at the beginning. The key is to stress to students the value of talking about the text in order to logically solve a problem. First, it is important to show students a variety of problems and have them successfully navigate identifying what the problem is asking them to accomplish. Without this understanding, solving accurately is difficult. Subsequently, modeling and practicing walking through the other steps of finding appropriate and relevant information, making connections to solve, and ultimately, applying knowledge to the broader mathematics world. The last step, being the hardest, may ultimately be used as an extra challenge for students who are ready to make those greater, broader connections.

Adaptations

Given the similarity of this strategy to the Mathematics Process Log, the recommended adaptations are also similar. When adapting the text-dependent questioning for students who are struggling with the literacy aspect of the strategy, provide sentence starters (as shown in the figure 3.3) to prompt student thinking. Also,

Task to solve: A Ferris wheel has a 100-foot radius and a 10-foot loading platform. As Theo and Jennifer ride the Ferris wheel, their height off the ground depends on the time they have been in their seat. Suppose it takes 12 minutes for their seat to make one revolution.

1. What is the maximum height off the ground for the ride?
2. At what times will Theo and Jennifer be 110 feet off the ground?
3. Write an equation of cosine that models this graph.

Mathematics text-dependent questioning for Ferris wheel problem ______________________

Text-Dependent Questioning	Questioning the Problem: General	My Focused Questions
Said what?	• What is the problem saying? • What is the problem telling me? • What does the problem say I need to clarify? What can I do to clarify what the problem asks? What does the problem assume I already know? • What is the problem asking me to do?	Radius loading platform time for 1 revolution
Did what? What information does the problem provide?	• What types of information are given [number, statistic, description, relationship, or visual]? • How does vocabulary give clues to solve the problem? • How does the text structure show mathematical patterns, connections, or relationships? • How does the problem indicate what it is asking to be solved? • How can I identify and efficiently and appropriately solve the problem? • What concepts, formulas, and ideas does the problem explain or reveal?	Use trigonometric graphs Use unit circle 100-foot radius and 10-foot loading platform Revolution = 12 minutes
So What? What connections can you make?	• Can I solve the problem? • Can I verify that my solution is accurate? • What does the problem want me to understand? • What connection is the problem revealing? • What is the problem arguing, and what support (evidence and reasoning) does the problem present? • How can I represent the solution (graphically, algebraically, numerically)? • Does my answer make sense?	Give amplitude by radius loading platform → minimum revolution told directly midline half of radius
Now what? Now, what can you do with your understanding of the problem?	• How does the problem connect with your previous knowledge or experience? • How does the problem influence or change your thinking? • What implications can you draw from the solutions to the problem? • Do more efficient options exist for solving the problem? • How does this relate to the learning targets for the course?	

Source: Eric Anderson, 2017. Used with permission.

Figure 3.3: Text-dependent questioning strategy and graphic organizer.

*Visit **go.SolutionTree.com/literacy** for a free reproducible version of this figure.*

listening to themselves and others articulate their thoughts out loud may help students better understand confusing ideas or concepts and clarify different thought processes. "Students become much more successful at higher order comprehension and monitoring their comprehension as a result of participating in questioning" (Duke & Pearson, 2002, p. 231). This shows how significant learning is when the students think aloud and question the why.

For students learning English, it may be helpful to focus on vocabulary in context so students have words to articulate their thinking. Teachers can provide a word bank of important or challenging words that are essential to explain the steps for solving problems. This is also a place where specific vocabulary strategies would be appropriate (word walls, Frayer models, and so on). See chapter 4 (page 63) for more vocabulary ideas.

As mentioned earlier, the fourth step in this graphic organizer can be difficult for many students. For that reason, it may be used as an extension activity for students who consistently and accurately provide evidence of proficiency with the first three sections of the text-dependent questioning graphic organizer. Also, as students become proficient with this process, teachers can reduce the scaffolding and have students simply work through four basic questions without all the prompting questions.

Considerations When Students Struggle

The goal of the two strategies in this chapter and the other strategies covered in upcoming chapters is to ultimately get students thinking like mathematicians. However, that may be a struggle for some students. Here are some questions to consider when you encounter students who are having a hard time applying the thinking process strategies in this chapter.

- **How might you ensure students are taking their time to engage in all the steps in the thinking process?** To answer this question, it is helpful to closely monitor student completion of graphic organizers. In the beginning, having students work through the steps in class will provide you with some observational feedback about whether students are taking their time or rushing through steps. Later on, monitoring will benefit teachers and students because it will allow you to see where individual students are proficient and where students may struggle.

- **How might you encourage students to use content-specific vocabulary when writing and speaking about the problem-solving process?** Mathematical vocabulary is incredibly important to thinking like a mathematician and understanding the problem-solving process. We encourage teachers and teams to build mathematics-specific vocabulary proficiency. We offer some specific vocabulary strategies in chapter 4 (page 63).
- **How might you help students make connections between various types of problems and problem-solving processes?** Mathematicians will tell you that mathematics is rarely a discipline with isolated skills. To be successful with mathematics, students must remember what they have learned and make connections to different concepts and mathematics units. It is important for mathematics instructors to help students recall related learning and help students make connections to other mathematical concepts throughout the year.

Look back at the two graphic organizers in this chapter. Where do you think your students may struggle? How will you help them through the parts you anticipate will be tough?

Considerations When Students Are Proficient

For students who have demonstrated proficiency, teams should collaborate on ways they can help students build greater connections between mathematical concepts while also helping students realize there may be multiple pathways to solving problems. The following are some other considerations.

- **Encourage students to share their problem-solving processes with others:** As we previously stated, there are often multiple ways to solve a problem and by having students share with each other, everyone benefits.
- **Consider using a recording program or app:** A recording program or app such as Explain Everything (https://explaineverything.com) is useful for creating a learning bank where students can watch and listen to peers solve problems to get help, study, and practice.

- **Assign proficient problem solvers to work with nonproficient problem solvers during practice time:** Not only do the proficient students reinforce skills by mentoring others, but the struggling students also benefit from the mentoring.

Every student, no matter their existing proficiency level, brings something unique to every mathematics problem, and teachers should celebrate those differences in their classrooms. Students have varied experiences and skill levels, so sharing their unique pathways to problem solving has value.

Wrapping Up

Recall that as mathematics experts, teachers will want to review their own thinking processes and then model them for students in order to help students think like mathematicians. Using the two strategies in this chapter will help students develop a logical process for solving problems and mathematical thinking. Eventually, students should be able to take their own steps to personalize their thinking processes.

Collaborative Considerations *for* Teams

- Which strategies are best suited for your thinking and your students?
- What do students need to think about before solving a problem?
- What do students need to do during solving?
- What should students be doing with the knowledge and solutions after solving?
- How can you differentiate these strategies and process steps for the different learners in your classes?
- How can your team work together to create, use, and model strategies that best teach all students?
- Which thinking steps connect to required mathematical common core principles?
- Does the team have access to student data that will inform it as to which thinking steps need to be reviewed, retaught, or reinforced?

CHAPTER 4

Prereading Strategies

As we wrote in chapter 3 (page 49), our collaborative team's journey began with the goal of teaching students to think like mathematicians. Prereading is significant because it helps to set up the learning process for students and supports their ability to think like mathematicians. In our experience working with mathematics teams, most mathematics teachers engage students in some sort of prereading activity before exploring a new concept or solving a basic problem or higher-level multi-step problem; however, we find the amount of prereading and the consistency vary depending on the level of complexity of the problems teachers are asking students to perform. Prereading was not always a priority for us on our mathematics team when assigning simpler one-step problems. The more conversations we had about prereading, the more we realized its importance as problems rise in difficulty and involve multiple steps.

This chapter explains the need for prereading in mathematics classes. After discussing the rationale for mathematics prereading instruction and the need for collaboration when combining literacy and mathematical expertise, we examine the four Ps of prereading for mathematics tasks. We highlight the prereading literacy skills embedded within the six CCSS Mathematical Practices (CCSS & NGA, 2010), how classroom setup can be very impactful for prereading instruction in mathematics, and applicable instructional strategies developed through collaborations with the mathematics teachers on our collaborative team. The chapter concludes with considerations for addressing students who continue to struggle and students who reach proficiency and are ready to take their learning farther.

The Importance of Prereading

One thing that students, and sometimes even mathematics teachers, forget is that prereading is an essential part of the reading process (Urquhart & Frazee, 2012). It frames how to approach a specific mathematical text so that students can interpret text, analyze concepts and principles, and ultimately solve a problem. Prereading sets the stage and prepares students' minds to take in and negotiate new information. This strategy is one that students can approach in several ways. On one level, prereading can be as simple as surveying and previewing a text or problem to gain an understanding of the content, scope, and depth. Further, prereading strategies may be more creative to engage readers and spark their thinking about how to solve a problem. Prereading strategies may also be more pointed and specific, clearly establishing a purpose for a specific problem or identifying what students are expected to learn from the problem. In grades 6–12, prereading strategies are important for both beginning mathematics students and more proficient mathematics students because they help students frame their commitment to learn from the text or problem, be it simple or complex. Prereading strategies prepare students for the reading and mathematics experience and what they will encounter once they are engaged in a specific type of problem.

From your teacher preparation classes, you might remember the concept of *schema*, the organizational network in the brain that takes in, stores, and connects information to other ideas (Rumelhart, 1980). It is not unreasonable for teachers to instruct their students about schema and make the concept understandable and relevant to them. Sometimes, in our work as literacy coaches, we refer to schema as *brain glue*: if students want information to stick, they have to get in the habit of activating their brain glue.

To reinforce this concept, you can regularly model and share your own mathematics and reading processes with students throughout the school year—for example, by relating a mathematical problem to a daily task they may encounter. For instance, how would someone use mathematics when picking out new furniture and trying to see what fits or doesn't fit into a certain space with unique dimensions? The teacher would model the process by guiding the students through each step in this real-world problem. When you regularly model the concepts, you can expect prereading activities to activate brain glue as students begin to build reading and mathematics behaviors and become more proficient mathematicians. When modeling, perform for students exactly what they will be doing when they work independently. Narrate the thinking process, including challenges they might face and how to overcome them.

Prereading strategies also help to activate prior knowledge and experience to help the reader make connections to new ideas or new information, and they work to prepare the reader in ways that help them focus with greater intention and deepen comprehension and understanding (Ferlazzo & Sypnieski, 2018). This is important in mathematics as most concepts and ideas build on each other. Students must recall previously taught information and concepts to solve problems at the next level. This is especially important in mathematics when solving problems with similar steps to work done previously. For example, students need to recall solving algebraic concepts when applying and solving geometric concepts.

Collaboration on Prereading Activities

When experts collaborate, they learn from one another and consider innovative ways to approach student learning. As literacy experts, we've found that many teachers from all subject areas have some knowledge of basic and intermediary literacy skills, like the ideas we discussed in chapter 1 (page 19), but we've found that many teachers from all subject areas need support in how to adapt and develop literacy strategies that are specifically tailored to their subject area and the difficulty level of their disciplinary texts. For instance, in teaching students to be more effective and efficient mathematics readers, reading specialists need to collaborate with mathematics teachers to think about how mathematicians read and think to adapt basic strategies and skills to fit their literacy needs. Specialists need to share their literacy expertise with mathematics teachers to build smarter, more innovative ways to think about literacy strategies specific to the mathematics classroom. To do this, our literacy team focused on how we could collaborate more effectively with our mathematics teachers.

Before we could focus on prereading strategies, however, we had to first consider the entire literacy-building process. Our first collaborative step was to interview our mathematics teachers to learn more from them about their reading and problem-solving process. We asked mathematics teachers questions like the following:

- "How would you read, approach, or solve this problem?"
- "What would you have to know to understand the reading or problem?"
- "How would you annotate or organize the information in the problem?"

We wanted to know more about how teachers approached the reading and understanding of mathematics, how they learned from reading about mathematics, and what they did after they finished reading a mathematical text or problem. From them, we gathered important insights about how they approached the reading and problem-solving process in ways that could direct our collaboration and decision making about prereading (discussed in this chapter), during-reading (discussed in the next chapter), and postreading strategies (found in this chapter and the next). From this process, we learned three valuable lessons.

1. We learned that mathematics teachers often previewed their reading of a problem before they started to solve the problem, gathering a sense of what the problem was about and what steps they would need to solve the problem. Previewing is a prereading strategy, which we explore throughout the rest of this chapter, that motivates mathematicians to read and process the problem.
2. Mathematics teachers shared that they need students to understand what information is pertinent in a problem and what information isn't needed to solve the problem. This is a tough skill and encompasses many processes at once since mathematics is recursive, as is reading. Thus, this lesson applies to both the prereading and during-reading processes. Prereading and during-reading skills are essential to the mathematics classroom.
3. Mathematics teachers often group and sequence mathematical information in a way that synthesizes concepts from the problem students will be reading and solving by underlining concepts and ideas as they read a problem for students. Students implement the underlined concepts when solving the problem (during reading) and also when understanding what the problem means and how the answer fits into the larger picture (postreading). During reading is discussed in more detail in chapter 5 (page 81).

We learned a valuable lesson from taking the time to discuss the mathematical reading process, and we began to understand and monitor some of the reading habits mathematics teachers wanted to teach our mathematics students. In doing this, we helped mathematics teachers recognize the value of approaching reading in strategic ways, and we were better able to design meetings to extend these discussions in more specific ways.

At the conclusion of this first collaborative step, we made it a priority to consider how we could focus on prereading strategies that could help establish a purpose for reading a mathematical text or problem and ways teachers might engage students prior to approaching a problem.

The Four Ps of Mathematics Prereading

When implementing literacy strategies, four prereading strategies are particularly important to consider: (1) *previewing*—scanning a text to gain an overview, (2) *predicting*—considering what the reading will say and what information it might provide, (3) *prior knowledge*—identifying what the reader might already know about a topic, and (4) *purpose*—establishing a focused reason for reading. The four Ps are active during the entire mathematics literacy process. Specifically, students go back to what the text tells them in the previewing section to predict what they think will come next or what step they should take, and then link that with their own knowledge of what the problem is asking, and finally establish a reason for what is being asked or what is being solved for. When students think like mathematicians, they are constantly evaluating the four Ps during prereading.

As you may have already observed, these four skills are not just for prereading: they permeate the reading process. Skilled readers constantly predict and connect to prior knowledge; less frequently, they preview and review the purpose during the reading. The important thing to remember is that if readers are to navigate texts successfully, they must learn and use prereading skills in a variety of ways to prepare themselves to enter the process of reading more successfully. Here are a few points to consider.

- **Model how to use prereading strategies with students:** Modeling aloud can help students understand how to use prereading strategies intentionally when approaching simple texts and problems or more complex readings and problems.
- **Focus on one prereading strategy with students:** While the reading process should be deliberate, no one wants it to become cumbersome, so teach students tactics for determining how the prereading strategy will best fit a particular task. The goal is to use prereading strategies flexibly and effectively.
- **Prereading strategies should become habits that don't take excessive amounts of time:** Although it takes practice to achieve proficiency, prereading strategies are designed to help the reader frame a purpose for reading and determine a motivation to read. The real work of reading to learn occurs during the next step in the reading process.

As your team of literacy experts works with your team of mathematics teachers to create prereading strategies, use the following as a basic guide to the four Ps of prereading that teachers can share with students.

- **Preview:** Students scan the reading or problem for various purposes: to review key words, examine images, read captions, or examine graphs they might need to understand. A good preview gives readers an idea of the content and problem they are about to tackle. It may encourage readers to think about what they already know about the topic or subject. It may lead readers to consider what the text or problem is about, and it may help readers clarify and make connections.
- **Predict:** Students think about what they might encounter or need to do to solve a problem as they make their way through the new text. Anticipating what is to come next in a text or problem is a good way to keep readers engaged and allow strategic readers to confirm or deny their thinking. This can lead to better comprehension and retention of the processes and steps in solving a problem. When using a mathematical text, you might ask students to predict or estimate what they think the answer or solution will be.
- **Prior knowledge:** David E. Rumelhart (1980) posits that all of our generic knowledge is embedded in our schema. More recently, experts have argued that by accessing prior knowledge, students are more equipped to tackle complex problems (Goodwin, 2017). By actively recalling knowledge, readers are prepared to build on the knowledge they already have and store the new knowledge for later use. By recalling what they know already, students are better prepared to attack a more complicated reading or problem and later connect it to developing schemas. Also worth noting is that if students have rich background knowledge or experiences in mathematical concepts, they are generally better prepared to handle more complex material to add it to prior knowledge. If students have little prior knowledge, be mindful of the difficulty level of the reading so they are not quickly overwhelmed.
- **Purpose:** There are two aspects of purpose that are important for readers to understand. First, readers should have a clear purpose for *why* they are reading or solving something. Based on our personal experiences, if a student's reason for reading or solving a problem is "because I have to," the student is probably not going to get a lot out of

the work. However, if a student can articulate the purpose for reading or solving the problem, odds are they will be more fully engaged and benefit from the experience more deeply. Second, students are more fully engaged with the text when they can *predict* and *anticipate* the author's purpose for writing the problem and what the author is looking for when the problem is solved.

Think back to the last problem you solved in your classroom. As a mathematics teacher and expert, what did you do before diving into the problem? What would you think about before solving it yourself?

- Do you notice yourself gravitating to any of the four Ps?
- What is most important for a mathematician to think about before solving the problem?

CCSS for Mathematics Connections

As stated previously, six of the eight CCSS Mathematical Practices (NGA & CCSSO, 2010) have a significant literacy component. We have chosen to highlight these practices because we've geared this text toward grades 6–12 mathematics teachers, and the complexity of each standard varies in different grade bands. Following are the six standards with literacy connections for review (NGA & CCSSO, 2010).

- **CCSS.MATH.PRACTICE.MP1: "Make sense of problems and persevere in solving them"** (p. 6). This standard requires an inner dialogue where mathematics-literate students explain to themselves the meaning of a problem, analyze the parts of a problem, explain components of and relationships within problems, and metacognitively monitor their progress toward a solution.
- **CCSS.MATH.PRACTICE.MP2: "Reason abstractly and quantitatively"** (p. 6). This standard requires students to conceptualize relationships between mathematics symbols and meaning.
- **CCSS.MATH.PRACTICE.MP3: "Construct viable arguments and critique the reasoning of others"** (p. 6). Not only are students required to build, analyze, and compare viable arguments and solutions

for problems, but they must also be able to look at the arguments and solutions of others and explain why those are or are not viable.

- **CCSS.MATH.PRACTICE.MP4: "Model with mathematics"** (p. 7). Mathematics-literate students can use mathematical concepts to solve real-life problems and models. This means students have to design problems, often by writing them out with mathematical symbols and language, and then solve and use the application.
- **CCSS.MATH.PRACTICE.MP5: "Use appropriate tools strategically"** (p. 7). Students must be literate enough to read a problem deeply and decide what materials are best suited to solve it—from paper and pencil to technology—depending on the problem at hand.
- **CCSS.MATH.PRACTICE.MP6: "Attend to precision"** (p. 7). Students are required to communicate precisely. They also are asked to precisely examine claims and use precise mathematics terms and definitions.

In the following section, we include a table (table 4.1) to show which prereading strategies teachers might use to address these CCSS Mathematical Practices.

Strategies for Supporting Students in Prereading

In our work as literacy coaches, we have examined past practices teachers use to introduce reading in the classroom. We examine how they might personally use prereading strategies when they read, and we explore other prereading options that might help to support their work with students in new, more innovative ways. For many mathematics teachers, this examination is a powerful moment because it's when they recognize the difference between what they do in the classroom, what they do personally as readers and mathematicians, and what they *could* ask students to do when reading and solving problems for the class.

Consider this: you know how you read and solve mathematical problems, and you're an expert at it. But it's easy to forget that you have to teach students how experts read and solve problems as well. Therefore, one of the most powerful adjustments you can make as a content-area teacher is to begin learning about and creating prereading activities that guide students to become strategic prereaders and solvers of mathematical problems. Many different teachers and teacher teams we have worked with have had a hand in creating the strategies and lessons for classroom use that appear in this section. Table 4.1 connects each of the strategies with the CCSS

Mathematical Practices that it supports. Each strategy combines the knowledge and experience of both literacy experts and content-area teachers, is explained, and includes adaptations that allow for differentiation for students who might struggle as well as for students who show proficiency. We hope the strategies—the Frayer model for vocabulary, the pod classroom structure, predicting and confirming activity (PACA), and vocabulary-concept memory game—spark ideas for you and your team to use and adapt for your own classrooms.

Table 4.1: Prereading Strategies and the Mathematical Practices

Strategies	MP1	MP2	MP3	MP4	MP5	MP6
Frayer model for vocabulary	x	x	x	x	x	x
Pod classroom structure				x	x	x
Predicting and confirming activity (PACA)	x	x	x	x	x	x
Vocabulary-concept memory game	x	x	x	x	x	x

Frontloading Vocabulary With the Frayer Model

When we meet with mathematics teachers, they often tell us that students struggle with knowing mathematical vocabulary because it is too challenging; concepts and vocabulary have different meanings in mathematics than they might have in an ELA text or science text. Students may struggle to understand mathematical concepts or definitions, and they may lack the ability to break down a vocabulary word into different parts. This makes understanding and interacting with the vocabulary students will find in the text or problem an essential part of the prereading process.

The *Frayer model* (Buehl, 2017; Frayer, Frederick, & Klausmeier, 1969) is a simple, adaptable strategy designed to help students demonstrate their understanding of complex mathematical vocabulary in a simple, adaptable graphic organizer. This helps students adjust or better understand the concepts in isolation and then work them into the context of the problem.

How to Use

The Frayer model graphic organizer identifies a vocabulary word and then asks students to address four parts to aid them in comprehending that word: (1) what it is (a definition), (2) what it is not, (3) characteristics, and (4) examples and images. As we show in figure 4.2, the Frayer model is adaptable to fit a variety of objectives. (See page 131 for a blank reproducible version of this tool.) The graphic organizer format allows readers to deconstruct a vocabulary concept into smaller parts to gain a larger understanding of the word or mathematical concept. By breaking the word down into different parts, students can draw connections between different concepts and identify gaps in their understanding. It is often successful in a jigsaw manner, where individual students take on different words and then share their work in small groups (Aronson & Patnoe, 1997).

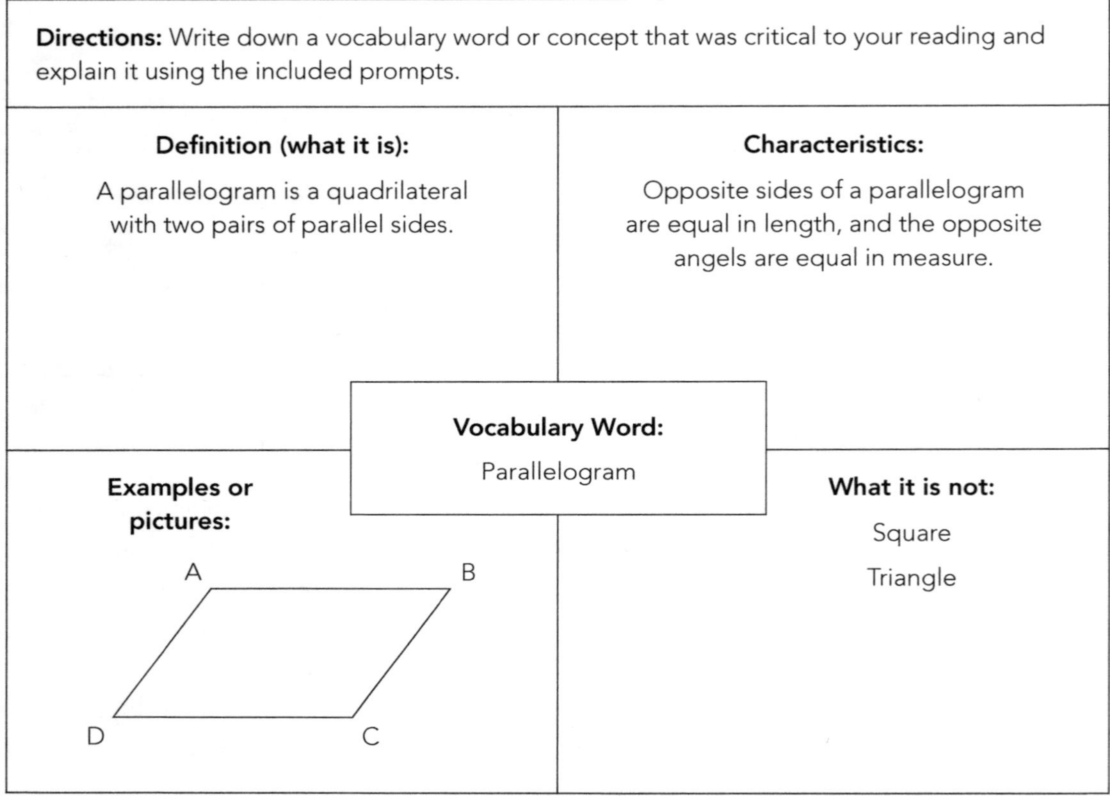

Figure 4.2: Example of the Frayer model graphic organizer for mathematics prereading.

Visit ***go.SolutionTree.com/literacy*** *for a free reproducible version of this figure.*

One way mathematics teachers can use the Frayer model for mathematical vocabulary is to have students skim through the learning targets (process standards) for a unit to find words they do not know. This also personalizes the learning for each individual by allowing students to choose what vocabulary they focus on based on their needs. After the prereading, our team of literacy experts and mathematics teachers developed a few short activities to have students share or check the knowledge they obtained and the words they learned. Figure 4.3 offers the steps for this process broken down into two days: a prereading activity and a follow-up postreading activity.

Day 1: Students Determine Important Vocabulary Words (Students should complete this independently after the teacher explains or models how to complete this activity.)

1. Students read targets from the unit of study and model thinking through the targets to focus their understanding of purpose.
2. The teacher models using the Frayer model graphic organizer by skimming the reading, problem, or targets. As part of this modeling, students should:
    - Read the definition. This information can come from the sources provided for them, the internet, class discussion, and so on.
    - Reflect on their understanding of the term if they understand how it is defined in the text. If one or more students do not understand, have them consider how they might identify other sources to clarify the meaning.
    - Fill in on the Frayer model a description of what it is not, based on the definition.
    - Skim the reading, problems, and targets again, and select characteristics that apply to the selected term.
    - Draw an example using the visuals in the text.
3. The teacher assigns homework based on what they feel is the next step in understanding the concepts.

Day 2: Students Engage in Optional Short Activities to Check Understanding and Address Targets (This is a postreading activity.)

1. The teacher has students compare graphic organizers for accuracy. (Students can use their specific graphic organizers from completing the activity in class the previous day or from completing it as homework.)
2. Students rank their Frayer model terms by specific criteria. (This is a collaborative activity with classmates, but the instructor should model it.)
3. The teacher has students compare and contrast various combinations of the terms in their graphic organizer. (This is a collaborative activity with classmates, but the instructor should model it.)

Figure 4.3: Process for applying the Frayer model.

Adaptations

You can adapt the Frayer model for students who are struggling to help them better understand the content. This strategy helps by creating a web of knowledge about a main idea or specific vocabulary words that could be challenging for these students to understand. Using the Frayer model helps students remember, access, and identify specific concepts; but for these groups, teachers should adapt the prompts in each box for accessibility. The answers and learning for *all* students should be the same. When you consistently use the strategy in this way, students struggling with mathematical terms become more familiar with the process and how to implement it to aid their understanding.

For students learning English, filling out the Frayer model boxes in both their primary language and in English can further assist with bridging gaps and providing a clearer understanding of the concepts. Another adaptation to consider when introducing the strategy is to provide a model for students and create a matching game using the vocabulary words you want to introduce and their corresponding definitions. Providing students with such a model outlines expectations for their work and allows mathematics students to interact more with the words.

In addition to adaptations for struggling students, some students may exceed the knowledge base and need a more challenging task. You can adapt the Frayer model strategy by changing the tasks within each of the boxes to demand higher levels of thinking. For example, the picture box could ask for pictures of what the vocabulary word is not, or a box can ask students to use the vocabulary word in a sentence. This is why the Frayer model is so adaptable for all students.

Organizing the Classroom With Pod Structure

Another prereading strategy is to set up your classroom to prioritize student collaboration. Placing students in groups (pods) is important to their success within the classroom environment as it provides for natural learning and collaboration among students while allowing them to learn from each other. This strategy involves grouping students by four and creates multiple pods.

How to Use

Students will benefit most from this strategy if teachers create pods with students of varying strengths and weaknesses; for example, one student is proficient in geometry proofs, two are moving toward proficiency, and one student is struggling with learning. The student who has achieved proficiency can take on the role of the leader, those who are working toward proficiency can learn from one another and the leader, and the student who is struggling will feel more comfortable asking questions and being part of the conversation in a small group to grow their skills. The small pod structure helps all learners in the group more readily participate in discussions and learn from each other. In the mathematics classroom, the pods may change and the roles may change depending on the unit being taught. All students have strengths and weaknesses, so pods change as student understanding of each concept changes. Each role provides an opportunity to hear what others are thinking and processing about a topic. This is powerful in the learning process of mathematics.

Adaptations

This structure is beneficial for all learners. English learners and special education students will all benefit from this structure since having peers close by to share their thoughts gives all students an entry point into the thinking. Teachers should be particular when placing students in the pods and be well aware of all students' needs. Students who understand the material immediately can help explain it to others. The students who struggle with the information will benefit from hearing the thoughts of their peers. Students in the middle have an opportunity to extend their thoughts and connect them with those of students who understand the material.

Predicting and Confirming Activity

The *predicting and confirming activity (PACA)* is another prereading strategy. This graphic organizer activity supports mathematics students to activate prior knowledge so students can make predictions about new learning they will encounter in a text or problem. In doing so, students set a clear purpose for their reading as they seek out information that confirms or refutes their predictions—important mathematical thinking skills. Although this is a prereading strategy, it lends itself to critical thinking throughout the entire reading process.

How to Use

For PACA, teachers encourage students to write their own predictions for the problem based on having previewed it. It may be helpful to preview the problem as a class and then allow students time to write predictions in the left column of the graphic organizer (see figure 4.4, page 76). While reading or solving (after making predictions), students write a plus if their prediction is confirmed, or they can write a minus if their prediction is refuted. In the right column, students record evidence from the text or problem that supports or refutes their predictions. Figure 4.4 shows a template of the graphic organizer. (See page 132 for a blank reproducible version of this tool.)

Problem:			
Prediction	**Confirmed (+)**	**Not Confirmed (–)**	**Support**

Figure 4.4: Sample predicting and confirming activity.

Visit ***go.SolutionTree.com/literacy*** *for a free reproducible version of this figure.*

Adaptations

For students with limited prior knowledge, it can be helpful to model how to develop predictions, walking students through each step and offering extra assistance or prompting. Using the think-aloud strategy shows the thought process of a teacher. Teachers may also provide a list students can reference to write their own predictions. This list can include terms, ideas, or equations about the content students are learning specific to different areas of mathematics, such as algebra, geometry, algebra 2, calculus, and so on.

Using a Vocabulary-Concept Memory Game

Vocabulary in isolation is important, but we also want students to make connections between different mathematical concepts and the ways in which those concepts work together to support larger understandings. This vocabulary-concept memory-game strategy is ideal to use before a lesson or near the conclusion of a unit of study because students have had ample time to work with the different core concepts, making them more likely to see how they are interconnected. In this way, students make use of all four Ps (previewing, predicting, prior knowledge, and purpose).

How to Use

Create game cards using important vocabulary and concepts from the reading (see figure 4.5). Content teachers and literacy experts can work collaboratively to decide these terms and concepts to ensure students are prepared to digest upcoming

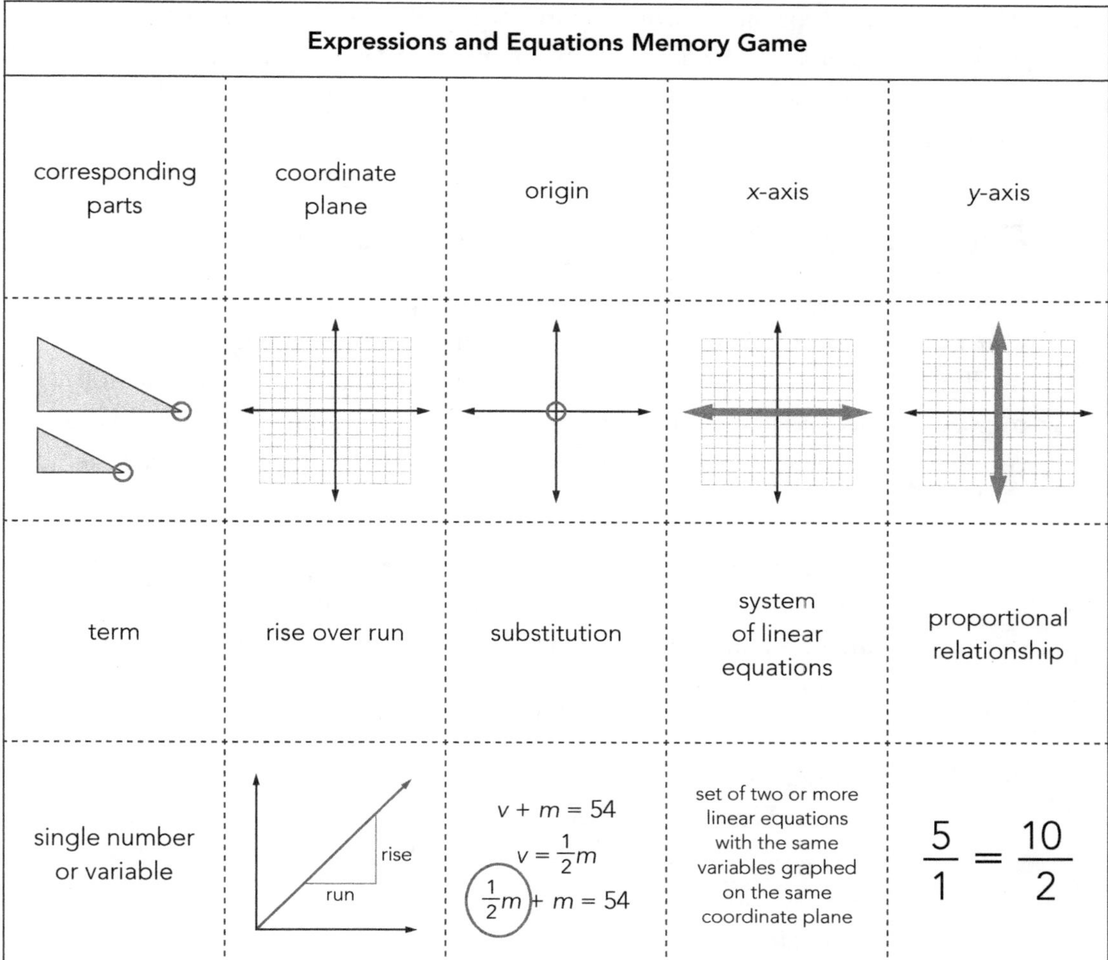

Figure 4.5: Vocabulary-concept memory game card examples and instructions.

reading or problems. In pairs or small groups, students take turns flipping over two cards. This strategy differs from the traditional memory game because, instead of looking for matches, the flipper has to build a connection between the two concepts shown on the cards. This can be a prediction, or students can do this after they have looked up vocabulary terms and defined them. In addition, students can write their connection between the two vocabulary words or concepts on a whiteboard. Through this process, students work together to link ideas and concepts.

Adaptations

To help students strengthen their vocabulary, add images and phonetic spellings to the cards so students better understand both the concept and the pronunciation of terms. The visuals can prompt or jog the memory for students of what the word or concept means while incorporating phonetic spellings that can be useful for students who understand the concepts but are still learning the language. Proficient readers may be able to handle more memory cards with additional vocabulary words.

Considerations When Students Struggle

Some students will struggle to use and master prereading strategies and the four Ps of prereading. Such struggles may occur because students are unable to recall prior concepts, they do not understand new concepts, or they are not able to make connections between concepts. Here are some additional questions to consider when you encounter students who are having a hard time applying prereading strategies.

- **How might you ensure students are regularly stopping and thinking before reading the mathematics problem or while reading the problem?** Should we use a symbol in our documents to signal them to stop and think about the problem? This symbol can be universal and used in all scenarios.
- **How might you provide students with additional prior or background knowledge they may need before reading the problem? What resources are available in your building to close knowledge gaps?** Is there a video of how the mathematical concept links to something in the practicality of everyday life? Can pictures help build some of the background knowledge?
- **How might you encourage students to make predictions before reading the problem?** Ask students to predict what mathematical steps they will need to take or what operations they need to use in each problem.
- **How might you help students understand why they are reading the mathematics problem and what they are reading for?** Provide additional space above or below the problem for students to write what they are solving for.

Note that your answers to these questions can be highly variable depending on the nature of your school, what you are working on with your students, and so on.

Review the strategies in this chapter. How might you help students who struggle to make progress?

Considerations When Students Are Proficient

For students who have demonstrated proficiency, teams should collaborate on prereading strategies that help students get to a deeper meaning of a text. Here are some additional ideas for proficient readers in the prereading process.

- **Vary the groupings so proficient readers are able to work with other proficient readers:** Encourage proficient readers to share their prior knowledge and make connections with other proficient readers and their prior information of the text or problem.
- **Allow proficient readers to work with nonproficient readers during the prereading process to share their understanding of the concepts (pod setup):** In helping relate their understanding of a problem, both proficient and nonproficient readers benefit from this collaborative work.
- **Provide extended choice opportunities to proficient readers (in addition to the choices you provide to all other students) that push them to higher thinking levels:** In implementing this practice, do not provide choice opportunities exclusively to highly proficient readers. The message should never be that only proficient readers are entitled to choice or additional learning opportunities.

You will note that for many of the strategies in this chapter, we acknowledge equal effectiveness for proficient and struggling readers. Every student, no matter their existing proficiency level, brings something unique to every text, and teachers should celebrate those differences in their classrooms. Students have such varied experiences and backgrounds that sharing their unique prior knowledge can create an exciting environment for learning while also showing students that their backgrounds and experiences, no matter what they are, have value. When there are opportunities to push proficient readers, teachers should do so. Bear in mind, though, that all students should have exposure to prereading instruction.

Wrapping Up

Using the prereading strategies in this chapter will help students activate their brain glue and be prepared to take in essential content information. You can implement all of these strategies in different classrooms and adapt them for all students to use.

Collaborative Considerations *for* Teams

- Which strategies are best suited for your students? Consider the following.
  - The needs of your students
  - The experiences and knowledge they bring to the reading
  - The experiences and knowledge they need to understand the reading
- Which strategies are best suited for the text?
- Which strategies are best suited for the targeted mathematics standards?

CHAPTER 5

During-Reading and Postreading Strategies

Kallie, a mathematics teacher working with ninth-grade students, noticed a troubling trend in several of her classes: after assigning word problems to her students, many students were returning to class confused about what information they needed to solve the problem and what information was not needed. Kallie knew the reading of word problems and the mathematical steps involved in solving them were critical to student success in her mathematics class, so she regularly planned for her instruction to support students' understanding of how to break down their understanding of mathematic word problems while reading. She did this by modeling her thinking using oral read-alouds and reteaching certain concepts as needed. She noticed that, as the year went on, more and more students had come to rely on her lectures and notes only for instruction and stopped attempting to reread the textbook as a reference or to access other resources when attempting to solve problems. Kallie knew that something had to change in her students' abilities to actively engage in the reading processes essential to learning mathematics confidently.

The prereading strategies explored in chapter 4 (page 63) begin the process for students to engage with their mathematics literacy, but that work is lost if students don't maintain that engagement in the work they do during reading. This chapter explores how you can work together with other mathematics teachers as a team to engage students during their reading and solving of mathematical problems. After looking at the need to build engagement and foster active reading, we review the six CCSS Mathematical Practices that contain literacy components (NGA & CCSSO, 2010) to highlight a series of during-reading literacy strategies we have developed through team collaborations with mathematics teachers. The chapter concludes with considerations for addressing students who continue to struggle and those who show higher-level proficiency.

The Need for Engaged, Active Readers

Try to remember the last time you assigned a complicated mathematical problem to your students that involved a lot of reading. Did they respond with enthusiastic high-fives and fist pumps, or did they shrug with a general sense of apathy? For many educators, the latter experience is the more likely student reaction when faced with a new and difficult mathematics task. The students most interested in the course content will surely get through the problem; other students may skim through the text to ensure they have a sense of what the problem is asking, and others rest their heads on their hands, quietly reading line by line but not truly engaging with the task. Most critically, as our story at the start of the chapter revealed, many students will fall short of completing tasks.

Imagine your quiet, compliant students reading the problem. Now compare this image with one of yourself at a time when you were fully engrossed in a mathematical text. When you become fully engaged while reading, you leave your current surroundings to enter the world of the text or problem. You tune out everything around you and begin to place yourself in the moment. Psychologist Mihaly Csikszentmihalyi's (2009) research reveals that this isn't limited to literary texts; reading an article about a new mathematical problem can just as easily pull us into a state of *flow*, a level of engagement where one becomes completely enveloped in a task or experience. Visualize your own body language when you are completely engaged in a text or problem and juxtapose that with the image of your students reading with weighted heads on hands. How do you make that image of your students look more like your own, including for those students who wouldn't list mathematics class on a list of things they're passionate about?

Linda B. Gambrell (2011), a former president of the International Reading Association, argues that students are motivated to read when they have opportunities to be successful with difficult texts. Gambrell (2011) also asserts when students have a clear sense of not only the *what* but the *why* behind a reading task, they are more likely to be engaged and find meaning in their work. Mathematics teachers have an obligation to provide students with mathematical texts that strike a balance between offering a challenge and not being overwhelming. When teachers ask students to make meaning with rigorous texts while also ensuring they have the tools they need to be successful, students take an increasingly active role in their learning.

In the mathematics discipline, being successful requires students to engage with a wide variety of text types: charts, graphs, data sets, images, textbooks, or word problems, to name a few. It also requires taking the time to activate prior knowledge and set a purpose per the prereading strategies in chapter 4 (page 63). Then, while the reading is actually happening, students need strategies to navigate rigorous mathematical problems successfully. Our hope is that the strategies in this chapter will help you and your students find the appropriate balance of rigor and success when engaging in mathematical reading activities. In taking these steps and using these strategies, you are already on the right track toward creating a more engaging, active reading experience for your students.

Consider your most recent mathematics assignment. How did you set it up for your students? Did you give them a specific task or problem? Did you simply tell them what steps to follow or engage in, or did you find helpful information to support their efforts? What were students expected to do while reading the mathematics problem you were asking them to solve?

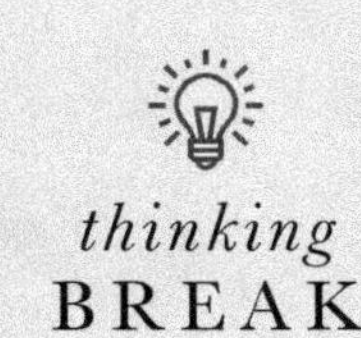

Collaboration Around During-Reading Activities

We have learned three important things from working with our mathematics colleagues about their reading habits, as we covered in chapter 4. The first centered on the value of prereading to mathematics readers. The second revealed how mathematicians frequently summarize and synthesize after reading mathematical texts (how they check their answers to see if they make sense). The third showed how mathematicians read for details while sorting and selecting key information and understanding what information is important and what isn't when solving a problem. Working with Kallie and other mathematics teachers helped guide our work toward developing strategies to increase student engagement while also developing essential during-reading skills in our students.

In each meeting with our mathematics team, we discussed matching desired content to a corresponding literacy skill best suited to help students access that content. This work required the team to address the first PLC critical question: What do we want students to know and be able to do? (DuFour et al., 2024). In matching content and literacy strategies, it was helpful that our team consisted of both mathematical content experts and literacy experts so that we could combine our expertise to find the common threads between what a student needed to know and the literacy skills they would need to learn it.

As literacy coaches, we had the opportunity to attend a meeting for a team of algebra teachers. They were examining formative assessment data, and team members noted that students were struggling to cite relevant evidence from a data table. Examining the assessment data helped us answer PLC question two: How will we know if they have learned it? (DuFour et al., 2024). The answers to the first two questions led to our collaboration about PLC questions three and four: What will we do when students haven't learned it?, and How will we extend the learning for those who have? (DuFour et al., 2024). The team recognized the problem but did not know how to teach students how to accurately cite data. Together, we explored different strategies that the team would commit to implementing in future instruction to help struggling students succeed and challenge those who had already demonstrated proficiency.

Replicating this approach with your own team will help you achieve the same success with students; however, where we had to identify or create instructional strategies to increase during-reading engagement, your team can reference and adapt from the strategies we include in this chapter. No matter how large or small your team may be, remember that this book is your literacy expert and the strategies in this book apply to many different mathematics courses. Use the knowledge and strategies in this chapter to make the connections between students' essential literacy skills and mathematics knowledge.

Remember that it is essential for teams in a PLC to collectively explore student data from formative assessments to identify patterns in learning. What does the data show about students' acquisition of skills and knowledge that help them achieve learning targets? What specific skills need more work or can be extended? Robert Garmston and Bruce Wellman (1998) advocate that teacher teams rely on protocols to help with the process of making these types of team decisions. Bill Ferriter's (2020) *The Big Book of Tools for Collaborative Teams in a PLC at Work* is also a tool to help teams build capacity to improve change. It focuses on the four critical questions of a PLC and can guide your team to address challenging questions. When teachers have a framework for dialogue about points of inquiry, they are allowed to explore their own assumptions and listen to their colleague's perspectives and concerns. This allows teams to make data-driven decisions that will have the greatest impact on their collective students' learning. Once your team has worked together to identify the gaps or opportunities for enrichment, it then has the tools it needs to choose one or two learning strategies most likely to address them.

CCSS for Mathematics Connections

As stated previously, six of the eight CCSS Mathematical Practices (NGA & CCSSO, 2010) have a significant literacy component. We have chosen to highlight these practices because we've geared this text toward grades 6–12 mathematics teachers, and the complexity of each standard varies in different grade bands. Following are the six standards with literacy connections for review (NGA & CCSSO, 2010).

- **CCSS.MATH.PRACTICE.MP1: "Make sense of problems and persevere in solving them"** (p. 6). This standard requires an inner dialogue where mathematics-literate students explain to themselves the meaning of a problem, analyze the parts of a problem, explain components of and relationships within problems, and metacognitively monitor their progress toward a solution.
- **CCSS.MATH.PRACTICE.MP2: "Reason abstractly and quantitatively"** (p. 6). This standard requires students to conceptualize relationships between mathematics symbols and meaning.
- **CCSS.MATH.PRACTICE.MP3: "Construct viable arguments and critique the reasoning of others"** (p. 6). Not only are students required to build, analyze, and compare viable arguments and solutions for problems, but they must also be able to look at the arguments and solutions of others and explain why those are or are not viable.
- **CCSS.MATH.PRACTICE.MP4: "Model with mathematics"** (p. 7). Mathematics-literate students can use mathematical concepts to solve real-life problems and models. This means students have to design problems, often by writing them out with mathematical symbols and language, and then solve and use the application.
- **CCSS.MATH.PRACTICE.MP5: "Use appropriate tools strategically"** (p. 7). Students must be literate enough to read a problem deeply and decide what materials are best suited to solve it—from paper and pencil to technology—depending on the problem at hand.
- **CCSS.MATH.PRACTICE.MP6: "Attend to precision"** (p. 7). Students are required to communicate precisely. They also are asked to precisely examine claims and use precise mathematics terms and definitions.

We include a chart to show which of three strategies—foldables and formula sheets, manipulatives, and word walls—teachers might use to address these CCSS Mathematical Practices.

Strategies for Supporting Students During Reading

When mathematics teachers take the time to establish a clear purpose for the reading and mathematics problem, students are already more equipped to understand the *why*. During-reading strategies help establish the *how*. This section explores during-reading strategies that both build on the prereading strategies in chapter 4 (page 63) and create a reading experience for students that is active and engaging.

The strategies in this chapter are designed to turn students from passive participants into active thinkers. Instead of merely getting through the reading or the mathematical problem, students will navigate their own way to new understanding. Along the way, these strategies encourage them to question and challenge a text or problem, embrace uncertainties, and engage in a dialogue with the problem. As part of teaching and modeling these strategies for students, you can—and should—share your passion for mathematics with your students. Connect that passion for the material to the process of learning it, and you will nurture and grow that same passion within your students. When you explicitly teach students to think like mathematicians, they are far more likely to find themselves engaged in new learning.

Table 5.1 connects each of the strategies in this chapter—word walls, manipulatives, foldables, and formula sheets—with the CCSS Mathematical Practices that it supports.

Table 5.1: During-Reading Strategies and the CCSS Mathematical Practices

Strategies	MP1	MP2	MP3	MP4	MP5	MP6
Word walls	x	x	x	x	x	x
Foldables and formula sheets	x	x	x	x	x	x
Manipulatives	x	x	x	x	x	x

Just like in chapter 4 (page 63), mathematics teachers working in collaboration with literacy coaches created or adapted all of the strategies in this chapter. You can use them to support instruction throughout grades 6–12 mathematics curricula. We accompany each strategy with an explanation and provide differentiation options for students learning English and students who qualify for special education as well as make recommendations for students who have mastered the learning. As with any strategy, it is important to consider ways to adapt these to suit your students and the learning outcomes you need them to achieve. Remember, good literacy strategies apply to simple and complex tasks. Our hope is that these methods become useful tools for your team and your students.

How can the use of prereading strategies help to establish a stronger sense of purpose for during-reading mathematical tasks?

Word Walls

A *word wall* is a collection of vocabulary words from the content, unit, chapter, or section of learning that are displayed around the classroom on a wall, bulletin board, or other display surface. Word walls are designed to be an interactive tool for students and contain an array of words that students can refer to and use during writing and reading. Word walls in the mathematics classroom serve as a visual reminder to help students recall specific content-area vocabulary they are currently learning or have learned. They provide a visual representation of the words, and support dialogue in the classroom, too. Word walls are a way to permanently model high-frequency words, encourage students to make use of academic vocabulary while they are collaborating, and they aid students in seeing patterns and relationships between words while also building their conceptual understanding of mathematical tasks. Likewise, word walls can provide support for learners during reading and writing assignments when you might ask them to explain how they problem solved.

How to Use

Making the words accessible on the walls in the classroom means that every student can benefit from the strategy. Write or type the words on a variety of background colors so students can easily read and understand them. This can be done easily on chart paper or a whiteboard. We suggest categorizing the words as they might relate to a unit or a mathematical standard that you are currently teaching. Try to make the word wall highly visible to all students and in a way that is easy to read and refer to throughout the unit. The idea is for student to be able to see the vocabulary they should be actively using when asked to do a variety of mathematical tasks independently, in collaboration with other students, or during whole-class instruction. Teachers and students should use word walls often, so students understand the importance of vocabulary in mathematics success. Being creative with word walls is what makes them unique to each mathematics classroom and engaging for students. One way of implementing this strategy is having students and teachers work together to determine which of the most commonly used words in the mathematics classroom and curriculum they should post on the word wall. The class can then add additional words gradually. We suggest limiting the amount of new vocabulary words to no more than five words per week—in ways that introduces new vocabulary regularly but does not overwhelm the learner. However, as the instructor, you are the expert in your class, and some classes might be able to handle more words per week. The word wall works best when teacher and students reference it daily and part of the classroom routine with multiple encounters. Some additional ways to use the word wall for vocabulary practice and reinforcement include chanting, snapping, cheering, clapping, tracing, playing word guessing games, and writing words within a classroom activity. These activities can help students get more familiar with saying mathematical words with ease, engage them as a classroom of learners, and remind them of the value of using academic language in ways to both understand and explain mathematical tasks.

Adaptations

Modeling the strategy is key, especially for students learning English and students who are struggling with proficiency. This includes modeling how to interact with the words. Students might not know what they don't know, or they might simply lack the ability to express mathematics processes using an unfamiliar language. Therefore, modeling identifies and states the expectations for the strategy. The more examples and modeling students experience the clearer they are about expectations.

For students who have previously demonstrated mastery with vocabulary, encourage and create opportunities for them to use the language they have learned frequently when answering questions and dialoguing in class. The more students use and express the words, the better the whole class understands the words.

Foldables and Formula Sheets

Foldables (graphic organizers) and formula sheets are the heart of an interactive mathematics classroom. They're helpful tools for students to use to organize new information and keep it accessible for reference. Foldables and formula sheets can look very different for each student or mathematics concept—there are many possibilities to implement their use in the mathematics classroom.

A *foldable* is a three-dimensional, student- or teacher-made interactive graphic organizer based on a skill. Making a foldable is a fast, kinesthetic activity for students that helps them organize and retain information either before, during, or after reading. The example in figure 5.2 is from a geometry classroom. The terms appear on the outside when the paper is folded. Opening each flap reveals the definition.

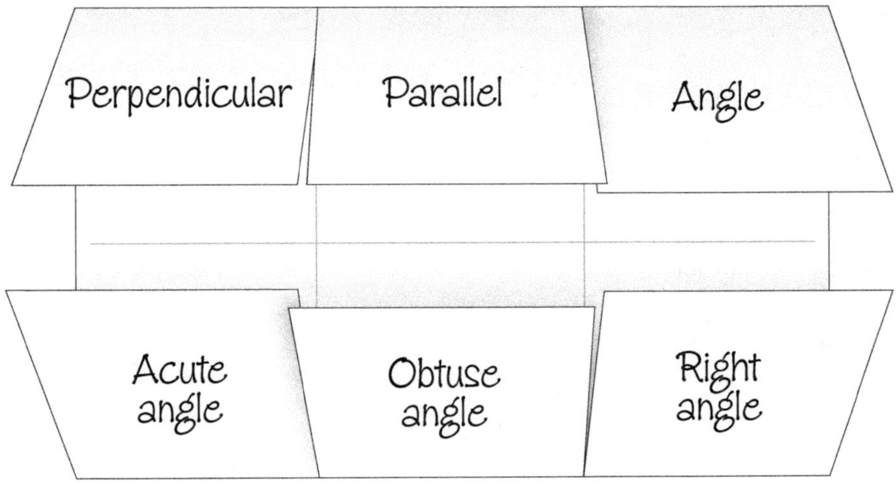

Figure 5.2: Example of a foldable for a geometry class.

A *formula sheet* is a progressive list or creative pattern of all the formulas students need to execute the mathematical problem. Formula sheets can be designed to look different for each unit or skill the class studies and organized in any way the teacher feels appropriate for the students to learn the material. Teachers and students should reference formula sheets as they are teaching or completing a problem. Formula sheets can be used in various capacities in pre-, during-, and postreading activities. Figure 5.3 (page 90) shows a formula sheet for a geometry class.

How to Use

Team members should work together to determine the important content in a mathematics text or problem and then what the foldable or formula sheet should look like. Teachers can then use the sample to model how to use both foldables and formula sheets. More importantly, be sure to show students how they can use both tools for their benefit. Will they be useful for a unit or new concept? Will they help students study for an exam? Are they a benefit that will reward their close reading and note-taking?

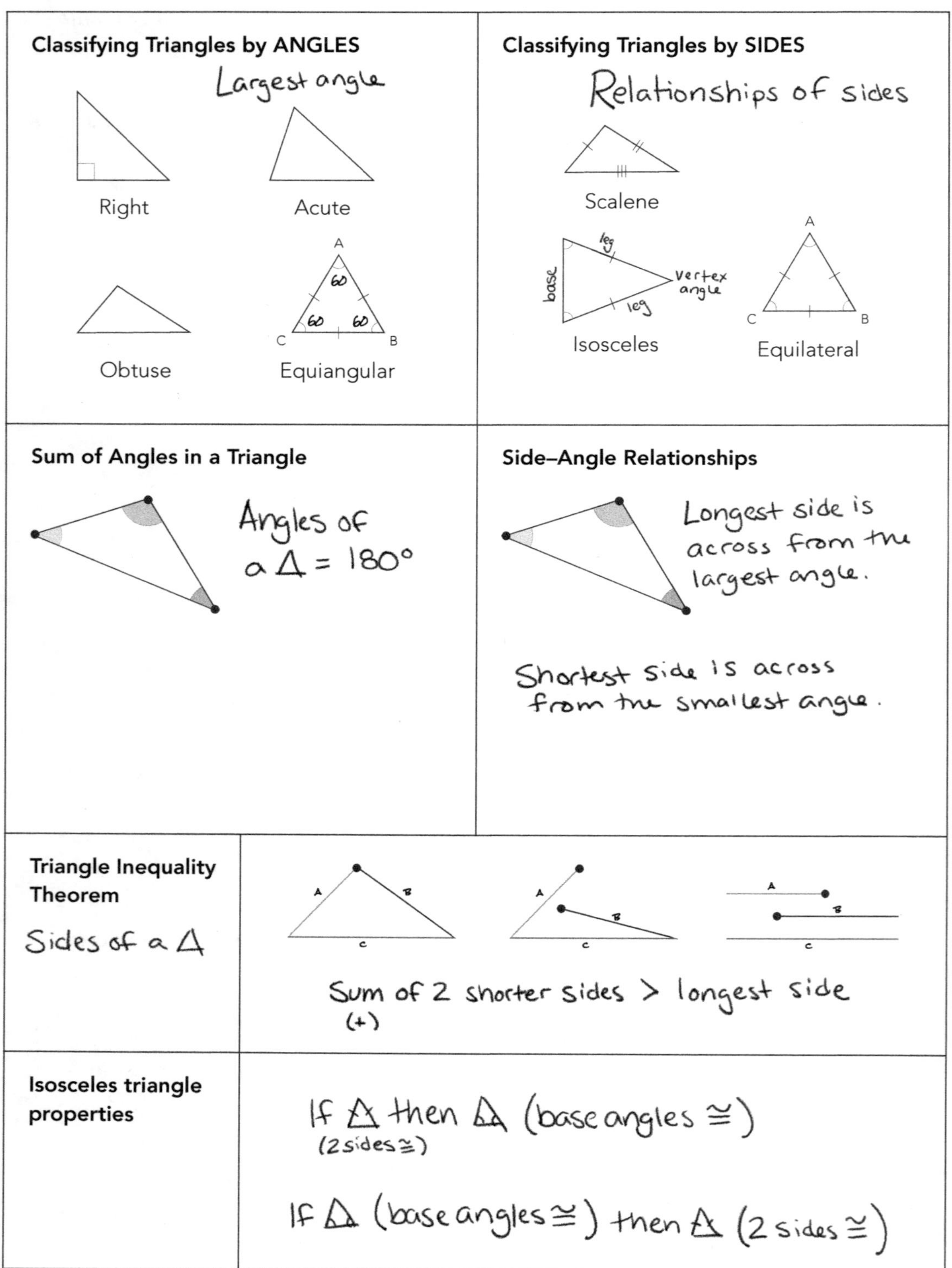

Figure 5.3: Example of a formula sheet for a geometry class.

As time progresses, encourage students to organize their foldable or formula sheet, including by choosing their preferred strategy. Encourage students to ask how they can use prereading skills to predict and set a purpose for their reading and then turn their foldables and formula sheets into logically organized charts or graphic structures that help them to organize their thinking while reading and doing the mathematics. This strategy requires just a quick explanation and provides students with simple, repeatable strategies.

Adaptations

Providing structure for students while they are working through new mathematical concepts and readings helps them understand what is being asked of them. Students learning English and students who are struggling with proficiency may need additional structure at first as they work to organize their foldable or formula sheet. While some students may be fine with one example or way to organize the foldables or formula sheets, other students may need more guidance with specific formulas and how to structure their tools.

Eventually, you want students to be able to do this work independently, but providing the extra structure at first can help narrow the focus and build the confidence students need to continue tackling the mathematical concepts and not give up. In addition, teachers can create an added space on the chart or organizer for struggling students to provide a general opinion or thought about the formulas. This can be added on the inside of the foldable if needed. This may help them make a personal connection or use a reminder to help them recognize a problem or formula.

For students who are proficient with knowing their formulas, allow them more flexibility in the structures of their foldables or formula sheets. A proficient mathematician is better equipped to articulate what is and is not working well for their understanding. That isn't to say any student shouldn't be able to make modifications or use the tool, but the proficient reader is more likely to identify where modifications are appropriate and how the tool can be used.

Using Manipulatives

Mathematics manipulatives are a common strategy to help students be more successful in the mathematics classroom at any age. *Manipulatives*—hands-on tools used to solve and examine problems—give students ways to construct physical representations of abstract concepts and ideas. Some might think using manipulatives is "too elementary" a strategy, but the reality is using manipulatives support students to make abstract ideas and concepts more concrete. This is true at any age and within any of the specific areas of mathematics. Manipulatives allow mathematics problems to come to life, so to speak. They help students construct an idea or understanding to connect vocabulary and symbols together. Some common manipulates in mathematics are blocks, spinners, shapes, counters, pieces of paper, and anything tactile for the students to touch and use to help represent the mathematics concept or problem. Using manipulatives builds students' confidence by helping them work though a problem as their use of the manipulative confirms or denies the accuracy of their thought process. When students have physical evidence of how their thinking works, their understanding is clearer. Using manipulatives also gives students more life experiences with an array of concepts and ways to solve problems. Manipulatives are used in many careers such as architecture, engineering, and in the medical field. In the same way, manipulative materials serve as concrete models for students to use to solve problems. This makes learning for the students more enjoyable and gives students a direct perspective of how mathematics links to the real world.

How to Use

Students learn better when they're engaged in their learning, and using manipulatives is one way to help engage them more fully within the classroom environment, especially the mathematics classroom. Following are many examples of how to use two types of manipulatives—foam dice and magnetic tiles—in a middle school or high school classroom. All ideas and suggestions can be adapted to meet the needs of your specific classroom and students.

- **Foam dice:** Link foam dice manipulatives to questioning or vocabulary to make the work more engaging. For example, use two dice—one marked with numbers 1–6 and the other with 7–12. The dice with 1–6 can be linked to unit questions and the dice with 7–12 to vocabulary questions. The physical dice and the suspense instantly make learning more engaging. Each student can roll the dice and a classmate can answer the question on the dice. The questions will be either about the content or about specific vocabulary. Another game to play is Fast Facts. This game is played with two opposing teams. Give the 1–6 die to one group and the 7–12 die to another group. A member from each team rolls a die, and the first player who shouts out the correct answer linking the concepts or vocabulary of the two dice wins a point. Once a team has ten points, they win and can start over. This strategy can be adapted to whatever concepts the class is learning.
- **Tile magnets:** Use colorful magnets with fractions on them during learning. The point of these manipulatives is that they can be moved

around and mixed and matched during learning to help students physically show their work. Get one of those big magnetic boards that also doubles as a whiteboard. When students finish their mathematics homework early, let them use the tiles to make fractions of shapes and use them to create ideas for the concepts they are learning. This is also a great way to teach many levels of geometry. The tiles also work well with fraction work. They can reinforce understanding of equivalent fractions, for example. Manipulative tiles make mathematical concepts more concrete and observable for students.

Adaptations

Using manipulatives is a great strategy for all learners but especially for multilingual learners and students with unique learning needs, such as mathematics learning deficits. Students with inferential comprehension and working memory goals can benefit from manipulatives. The sensory aspects of the manipulatives—color and tactile experience particularly—are important to the learning process for some students. All the games with manipulatives can be adapted to fit your students and curriculum. Colorful manipulatives help to differentiate ideas, numbers, or letters to support learning. Note, however, that students who have mastered learning targets or concepts may not find manipulatives helpful.

Postreading Consideration

The strategies discussed in this chapter can all support the postreading of mathematics text. In mathematics, the strategy is to help support student thought processes throughout the entire working process with a problem. The postreading in mathematics is reflecting on the question or problem and then asking, "Does the solution make sense?" If the solution doesn't make sense, students use the during-reading strategies to rework the problem. Postreading in mathematics looks different than in other disciplines, but the focus on reflection and seeing if the solution makes sense helps to clarify and put postreading into more simple terms.

Considerations When Students Struggle

Students who struggle with during-reading exercises will benefit from repetition (reteaching) and exposure to multiple reading strategies that engage their higher-level thinking skills. As with prereading, struggling during reading can occur for a variety of reasons: the mathematics itself may be too high a level and create frustration, some students may need additional instruction and practice with a specific strategy, or the vocabulary may be too technical or challenging. Every student is bound to encounter their own unique challenges. Here are some additional questions to consider when you encounter students who are having a hard time applying during-reading strategies.

- **How might you help students improve the focus of their annotations or notes when reading and engaging in a mathematical problem?** Can you provide them with a specific guide and outline for their notes? When the students become more comfortable with their annotations or notes, scale back the support.
- **How might you help students who have vocabulary deficits and difficulties?** Provide students with an enriched vocabulary classroom to increase their interaction with new words, which will ultimately help with their metacognition of new words.
- **How might you help students stay focused during a mathematical problem and understand what question is being asked?** Keep structure in the classroom instruction and guide students during the mathematical problem or question.

We believe that the strategies in this chapter provide practical instructional approaches for teams trying to increase student engagement while reading within mathematics. At the same time, teachers should solicit feedback (gather data) from students to find out what is working best and where they find the strategies to be less helpful. This sends the message that teachers value students' feedback, and seeking students' input provides them with a sense of ownership over their learning. Teachers can report that feedback to their collaborative team, and the team can use that information, along with other formative assessment data, to make informed decisions about the direction of instruction. The response may be to adapt an existing strategy that might have a greater impact on students' learning. It's important to remember that students need practice with strategies—it often takes teachers (in terms of instruction) and students (in terms of learning) multiple repetitions before they can master a strategy.

thinking BREAK

Review the strategies in this chapter. How might you help students who struggle to make progress?

Considerations When Students Are Proficient

For students who have demonstrated proficiency, teams should collaborate around adopting instructional approaches that are challenging and thought-provoking. In addition to the adaptations in this chapter's strategies, here are a few considerations for further differentiating instruction for proficient readers.

- **When grouping students so that proficient readers are working with other proficient readers, encourage them to share their thinking with their group while completing a mathematical problem:** This strategy challenges each student in the group to increase his or her own mathematics comprehension toolbox.
- **When varying groups to match proficient readers with nonproficient readers, encourage proficient readers to share their process for making meaning from texts:** This allows them to serve as mentors and leaders to other students in the classroom.
- **Develop reading tasks that are appropriately challenging:** Many proficient readers are risk averse. Pushing these students outside their comfort zones doesn't mean creating more work for them; rather, it ensures they will be challenged in ways that will encourage growth. Teacher teams should plan for ways that extend learning for any student that builds on a mathematics reading problem or activity.
- **When you embed into your during-reading strategies choices for students, ensure those choices are available to all students:** The implication should never be that only proficient students are entitled to choose. When students of all proficiency levels have choices, it sends a message to all students that they have a say in their learning.

During reading, teachers commonly hyperfocus on ensuring that still-developing students attain the minimum learning necessary for success on assessments and other postreading tasks. However, a school that functions as a PLC doesn't just want all students to learn; it wants all students to learn at *high* levels. Encourage proficient student mathematicians to stretch and break out of their comfort zone. Through peer mentoring, students reinforce their ideas and gain important confidence in working with others. Also, by providing times for all students to engage in additional mathematics problems or apply the concepts to real life and materials beyond the text—for interest and learning—proficient readers can move beyond the general learning targets while their peers choose problems that reinforce current targets. Ultimately, the more thinking teachers can encourage proficient students to undertake, the more they will solidify their knowledge and remain engaged in learning.

Wrapping Up

Thoughtfully planning during-reading activities allows students to practice comprehension and engagement. If you provide them with ample opportunities to practice and repeat strategies, students will internalize those strategies, ensuring they are strategic readers as they solve mathematical problems in future years and courses.

Collaborative Considerations *for* Teams

- Which strategies are best suited for your mathematics course instruction? What might work best for your students? Consider the following.
  - Do student data indicate specific strategies that would benefit differentiated groups of students?
  - How can the team supply background information and a purpose that engages students and helps them focus on mathematical concepts and understanding?
  - What supports do students need to focus on the important content in the mathematics reading?
- Which strategies are best suited for developing the understanding of mathematics and doing mathematics?

CHAPTER 6

Strategies for Writing About Mathematics

Kallie has continued to work with her students on being more independent and using prereading, during-reading, and postreading strategies to clarify their understanding of reading, thinking, and writing like a mathematician. This chapter highlights the ways mathematics teams and literacy experts can work together to make meaningful revisions to instruction in order to mentor students to think like mathematicians when writing for specific disciplinary purposes. First, we share how teachers we have worked with were able to enhance their writing instruction by collaborating with teachers in other disciplines. We stress the importance of a common schoolwide writing vocabulary aligned with the language used in the CCSS for Mathematics (NGA & CCSSO, 2010). The chapter concludes with concrete writing strategies that mathematics teams can implement to enhance literacy instruction in their classrooms.

The Need for Common Writing Terminology

On the surface, mathematics does not seem to be an educational field full of literacy, but the discipline has many opportunities to use literacy skills, including writing. Word problems, geometry proofs, solutions to problems, and problem-solving explanations all involve writing (and sometimes speaking) in modern mathematics classrooms. While some mathematics teachers do provide lessons on writing, often mathematics teachers rely on skills students already have been taught to write in their English classes and previous mathematics experiences. However, as students progress as writers through middle school and high school, the differences between English writing and mathematics writing increase. For example, our school realized that mathematics teachers were using different terminology with their writing requests and instruction.

Why was the terminology suddenly different? During the first two years of our literacy work, we sought interested volunteers from all content areas to share their thoughts about writing over lunch. The response was robust, as we found teachers from almost all content areas willing to share. Like all good teams, this group sought to solve a problem: How could we help students make more sense of writing across the curriculum?

The first meeting of this writing team consisted of teachers from different areas sharing their writing expectations, structures, and samples. Subsequent meetings involved teachers from different disciplines examining the different types of writing and finding commonalities. From those conversations, it became clear that as a school, one way we could problem solve would be to consistently use similar writing terminology such as *claim* (instead of *thesis*, *hypothesis*, and so on), *subclaim* (instead of *topic sentence*), *evidence* (instead of *quotes*, *examples*, *data*, and so on), *reasoning* (instead of *justification*, *elaboration*, and so on), and *conclusion* in all content area classes.

During these initial meetings, we found that mathematics teachers seldom participated because they did not see themselves as teachers of writing. However, a few teachers became interested in this common literacy vocabulary and experimented with teaching their students that answers to problems were similar to writing claims. Simply by using the word *claim* in their work, these mathematics teachers found students' answers to be better composed and clearer. In many ways, since success is contagious, this experiment with claims became the impetus for greater literacy work with our mathematics team.

Support for Common Language

Admittedly, our content-area teachers did not just come up with the terms *claim*, *evidence*, and *reasoning* on their own. Those terms are key elements of argumentation; they are embedded in the English Language Arts Common Core State Standards (NGA & CCSSO, 2010), and they are referenced in the CCSS for Mathematics (NGA & CCSSO, 2010).

One of the first times we encountered the common argumentation language was when we met with Katie S. McKnight in September 2014. She brought to our attention the work of former teacher Eileen Murphy and her colleagues at ThinkCERCA (www.thinkcerca.com; n.d.a), a literacy courseware company. *(CERCA* stands for *claim*, *evidence*, *reasoning*, *counterargument*, and *audience*.) Since that meeting, different groups of teachers and teams worked with that pedagogical terminology, and after a few years of the ideas organically filtering through different divisions and collaborative teams, our content area teachers decided to loosely adopt the terminology. According to ThinkCERCA (n.d.b):

> Teaching students how to make claims, support their claims with evidence, explain their reasoning, address counterarguments, and use audience-appropriate language is the most effective way to improve achievement on assessments and prepare students for post-secondary life.

With research and logic behind us, we were happy to facilitate the proliferation of this common language.

While the common language of ThinkCERCA has made a significant impact on our disciplinary teachers and students, we recommend that your teams work together to create your own set of common literacy terms that work for you and your school in order to empower your collaborative teams and help your students develop a connected sense of literacy across the curriculum. Students benefit when they hear similar writing vocabulary across content areas.

If CERCA terms do not make sense for your team, consider ways that you can collaborate with colleagues in different disciplines to gain a sense of the writing vocabulary used in your school. Who might be able to help you discover language used in different disciplines? How might you work together to construct a common language to use in your school—and who would help out?

How can you implement the ThinkCERCA (n.d.a) model or other argumentation tools in your classroom?

CCSS for Mathematics Connections

As we have stated in previous strategy chapters, six of the eight CCSS Mathematical Practices (NGA & CCSSO, 2010) have a significant literacy component. We have chosen to highlight these practices because we've geared this text toward grades 6–12 mathematics teachers, and the complexity of each standard varies in different grade bands. Following are the six standards with literacy connections for review (NGA & CCSSO, 2010).

- **CCSS.MATH.PRACTICE.MP1: "Make sense of problems and persevere in solving them"** (p. 6). This standard requires an inner dialogue where mathematics-literate students explain to themselves the meaning of a problem, analyze the parts of a problem, explain components of and relationships within problems, and metacognitively monitor their progress toward a solution.
- **CCSS.MATH.PRACTICE.MP2: "Reason abstractly and quantitatively"** (p. 6). This standard requires students to conceptualize relationships between mathematics symbols and meaning.
- **CCSS.MATH.PRACTICE.MP3: "Construct viable arguments and critique the reasoning of others"** (p. 6). Not only are students required to build, analyze, and compare viable arguments and solutions for problems, but they must also be able to look at the arguments and solutions of others and explain why those are or are not viable.
- **CCSS.MATH.PRACTICE.MP4: "Model with mathematics"** (p. 7). Mathematics-literate students can use mathematical concepts to solve real-life problems and models. This means students must design problems, often by writing them out with mathematical symbols and language, and then solve and use the application.
- **CCSS.MATH.PRACTICE.MP5: "Use appropriate tools strategically"** (p. 7). Students must be literate enough to read a problem deeply and decide what materials are best suited to solve it—from paper and pencil to technology—depending on the problem at hand.
- **CCSS.MATH.PRACTICE.MP6: "Attend to precision"** (p. 7). Students are required to communicate precisely. They also are asked to precisely examine claims and use precise mathematics terms and definitions.

Table 6.1 shows which CCSS Mathematical Practices can be addressed by each of the two strategies in this chapter.

Table 6.1: Writing Strategies and the CCSS Mathematical Practices

	MP1	MP2	MP3	MP4	MP5	MP6
Evaluating and supporting claims using evidence	x	x	x	x	x	
Using color-coded paragraph proofs	x	x	x	x		

Strategies for Creating Expert Writers

There are so many different ways to get students to produce thoughtful writing. More often than not, strategies come down to modeling and practice. Along with utilizing a common writing vocabulary, using mentor texts, creating models based on students' efforts, and cowriting can all lead to positive results over time. When possible, work with other teachers to share samples, practice ideas, problem solve, and provide feedback. When different teams within a school community come together to share the ways they teach and assess writing, they can make powerful connections to the ways that all students engage in writing tasks across different disciplines.

The following strategies—evaluating and supporting claims using evidence and using color-coded paragraph proofs—created in conjunction with our mathematics teachers, will help other mathematics teachers and teams build expert writing competencies with their students.

Evaluating and Supporting Claims Using Evidence

This strategy works best when connecting the skills to argumentative writing skills in other disciplines. Students should be familiar with writing claims or thesis statements in English class, social studies, and in other disciplines. Students know that these statements provide answers to questions asked by teachers. Students also know that they have to provide evidence to support their claim or thesis. For this strategy in the mathematics classroom, students are asked to evaluate answers to mathematical problems (claims) and then independently identify evidence from the actual problem, the calculation process, their knowledge of proofs, other resources, or both that support the correct answer or claim statement. This is a powerful formative assessment in that students are able to see a range of answers or claim statements, and then they are required to find evidence that best supports it. By scaffolding instruction to help mathematics students compose answers as an effective claim and then following that by helping students to identify the most appropriate evidence, teachers are able to break down the writing and thinking process into separate stages. When students lean into common language like *claim* or *thesis* and *evidence*, our mathematics team finds that familiarity can help students better communicate answers and achieve greater success on mathematical problems and assessments.

How to Use

Teachers provide students with a mathematics problem and three or more claim statements that could be answers to the problem. Students are then asked to evaluate the claim statements and select the correct answer. See figure 6.1 for an example from an algebra class. After students select their answer or claim, this may be a good opportunity to do a formative check and have a discussion about why students made specific selections. The alternative is to allow students to begin to select evidence for whichever claim they selected. If students choose the most effective claim, they should have little difficulty finding appropriate evidence. If students choose a less effective claim, they may struggle to find evidence, which may require them to rethink their original selection. Students are encouraged to identify multiple pieces of evidence and provide reasoning for each selection.

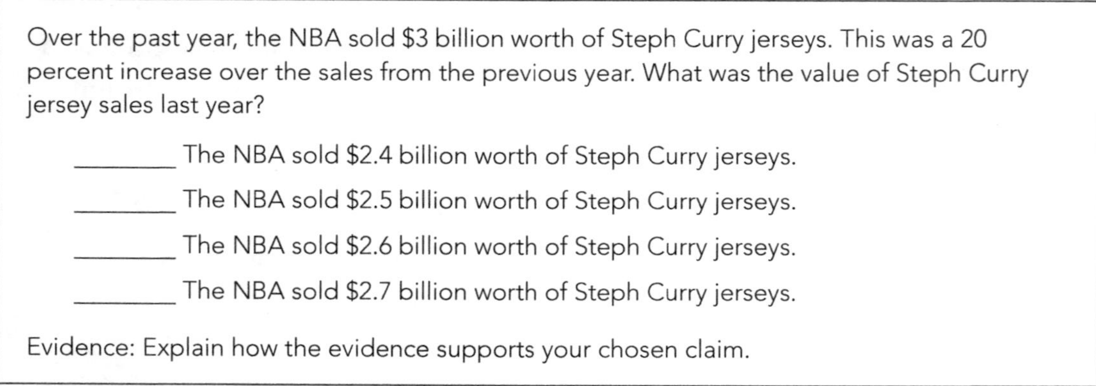

Over the past year, the NBA sold $3 billion worth of Steph Curry jerseys. This was a 20 percent increase over the sales from the previous year. What was the value of Steph Curry jersey sales last year?

________ The NBA sold $2.4 billion worth of Steph Curry jerseys.

________ The NBA sold $2.5 billion worth of Steph Curry jerseys.

________ The NBA sold $2.6 billion worth of Steph Curry jerseys.

________ The NBA sold $2.7 billion worth of Steph Curry jerseys.

Evidence: Explain how the evidence supports your chosen claim.

Figure 6.1: Algebra class evaluating and supporting claims using evidence example.

Adaptations

You can adapt this strategy for English learners and struggling students by guiding them to select the correct answer. Students who choose the most effective claim should have less difficulty finding appropriate evidence, but students who choose a less effective claim may struggle to find evidence, which may frustrate them.

Also, you can reverse this strategy for struggling students. First, provide the evidence for them and then have them come up with the answer or claim. This helps them to process the information and try to fit it all together without having to initially come up with the claim's evidence on their own. Eventually, students can use the strategy as previously described, but with scaffolding at first to help them work with the ideas.

For students who show proficiency and consistently choose accurate claims, teachers can adapt the strategy by having these students compose their own claims without a provided list.

Using Color-Coded Paragraph Proofs

In general, there are two ways to write geometry proofs in mathematics courses: the two-column method and the paragraph proof method. The latter can be used to build student writing skills while doing geometry.

When students are learning how to write paragraph proofs in mathematics class, they frequently ask, "How do I write it?" or "How much do I need to write?" Additionally, they want guidance to make sure that they have all the necessary components for a proof. *Color-coding drafts* of student proofs is an effective visual strategy to help students see how each component of the proof contributes to the larger whole. Students will have a clear visual indicator that they have included all proof components. If desired, by comparing their colored proofs with exemplar or peer writing, students will be able to assess their own writing for quantity. Students can attempt to assess their own or peers' work quality; however, teachers will ultimately have to check for correctness.

According to "What is a Paragraph Proof?" by Chelsea Coons (2022), geometric proofs have four components that can be connected to common literacy language in a school. "Givens" and "prior knowledge and geometric properties, theorems, and definitions" provide evidence for proofs (Coons, 2022). Students then justify the evidence through logical reasoning. Finally, students finish the paragraph with a claim or statement that proves the concept. By connecting these parts of the proof to a common language—evidence, reasoning, and claim—that students may encounter in other disciplines, students feel more comfortable putting their ideas into words.

How to Use

Assign a color for every required component of the writing task. That is, for proofs: the givens (evidence) are yellow; the prior knowledge, properties, theorems, and definitions (evidence) are green; the justification (reasoning) is blue; the claim or proof statement is red, and so on.

Students then use a highlighter or a word-processing application to highlight the entire paragraph, using the designated colors for every proof component in the paragraph. Students can compare their highlights with a peer or exemplar text to reflect on their progress. Even though there is no set length for proofs, patterns should emerge. From the highlights, students should be able to ascertain that they have the necessary components.

Figure 6.2 shows a sample problem (Coons, 2022) with the paragraph coded using shapes.

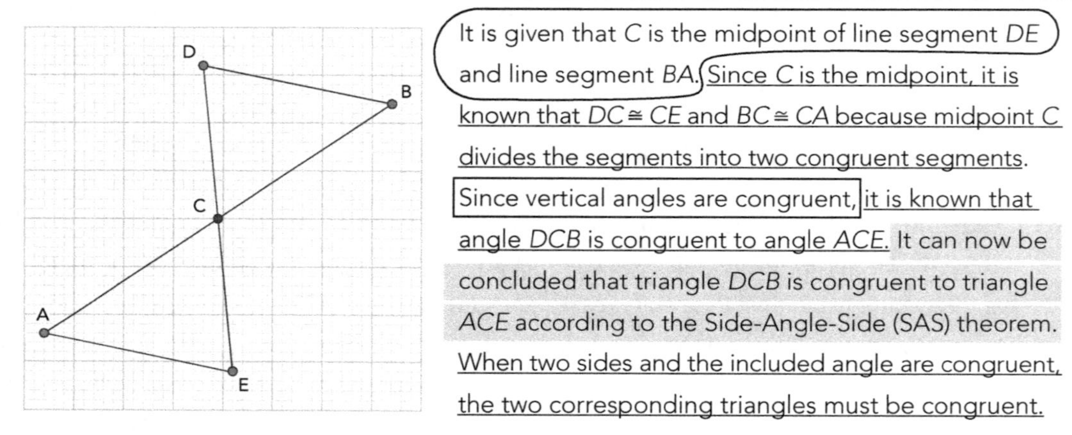

Source: Adapted from Coons, 2022.

Figure 6.2: Example of a geometry proof.

Adaptations

This visual strategy can be helpful to all students—including language learners and struggling students. Using the various colors visually gives the students a clearer idea of their writing and thought process. This also shows students that they have included all the required proof components. Students can use specific colors for all their writing. Visually this helps the students to understand what they are writing and make sure they have tackled all the required components of the writing task.

Considerations When Students Struggle

We realize that some students will struggle to use and master writing strategies in mathematics class. Such struggles may occur because students struggle with writing concepts like claim, evidence, and reasoning. Here are some additional questions to consider when you encounter students who are having a hard time applying writing strategies.

- How can you reinforce the concepts of claim, evidence, and reasoning for students? Are there English teachers who can help?
- How might you provide students with additional models to use when they write their own claims, evidence, reasoning, and proofs? What resources are available in your building to help provide models? Are there online resources that can provide extra examples?
- How might you encourage students to think through problems like a mathematician in order to make sure they include all of the elements of proofs? Are there reflection practices that students can engage in to check that they have included the required elements?

Note that your answers to these questions can be highly variable depending on the nature of your school, what you are working on with your students, and so on.

Review the strategies in this chapter. How might you help students who struggle to make progress?

Considerations When Students Are Proficient

For students who have demonstrated proficiency, teams should collaborate on writing strategies that help students increase the accuracy and depth of answers. Here are some additional ideas for proficient readers in the writing process.

- **Vary the groupings so proficient writers can work with other proficient writers:** Encourage proficient writers to share their work with other proficient writers to check for accuracy and depth.
- **Allow proficient writers to work with nonproficient writers during the writing process to share their understanding of the concepts (pod setup):** In helping relate their written answers, both proficient and nonproficient writers will benefit from the models and modeling they can provide for each other.

It is important to note that student success with writing in mathematics may be independent of student problem-solving skills. Proficient problem-solvers may struggle with writing skills and struggling problem-solvers may be able to write well. Teachers should be ready and willing to provide models and seek support from colleagues to help with writing skills that may be required for mathematics writing but may not be in every mathematics teacher's comfort zone.

Wrapping Up

Using these strategies will help students grow as written communicators in the mathematics classroom. It is good to keep in mind that in writing, imitation is the best form of flattery as well as a great way for students to develop their own skills and independence. That being said, it is also important to challenge proficient writers by asking them for more depth and accuracy, and potentially to help with modeling for other students.

Collaborative Considerations *for* Teams

- Which strategies are best suited for your students? Consider the following.
    - The written skills of your students
    - The experiences and knowledge they bring to the writing
    - The standards you want to teach and reinforce
- Do English teachers (and social studies, science, and so on) use the terms *claim*, *evidence*, *reasoning*, and *justification*? Or do they use other terms like *thesis* and *explanation*?
    - If not, we recommend divisions work together to consider common terminology to alleviate student confusion.
    - If they do, how can you leverage their terminology to help make things even clearer for your students?

CHAPTER 7

Assessment

Many of our team's literacy coaching sessions involve detailed, lengthy discussions on what exactly we are trying to assess. Are we assessing content knowledge? Or are we assessing literacy skills? Ultimately, the answer is both. If teams want students to grow and succeed, teachers need to hold each other accountable for ensuring all students receive feedback, which includes assessing their work and committing to not only careful and intentional content and skill instruction but also a strategic assessment that tackles these learning goals. Teachers must assess students' ability to understand the mathematics versus understanding the language mathematics is conveyed with. If teachers don't recognize when the data points toward one or the other, they won't be able to intervene effectively to ensure students reach proficiency.

During our sessions, we collaboratively set out to find ways to make most literacy assessments quick and seamless but still informative. Our goal was to ensure that disciplinary content was not lost while also ensuring assessment of proficiency with power (essential) standards and skill development related to literacy. We also wanted to ensure any growth of these skills transfers beyond the assessment.

Together, we found that through quick formative checks and feedback, we could assess student literacy work and produce disciplinary growth. Sometimes, the assessments came in the form of warm-up activities, like an entrance slip on a previously investigated concept or as a brief 3-2-1 response. Sometimes, the assessments were simple visual scans of whether students ultimately agree or disagree with a statement. The challenge (and the fun) of working as a collaborative team ensured that, from start to finish, activities were useful for both literacy and assessment purposes—and teachers felt comfortable implementing them in their classes not only because they created them together but because they knew the team would also work together to assess the resulting data and use it to further instruction.

There are seven considerations to keep in mind when approaching assessment, including (1) understanding the role of literacy-based assessment, (2) matching text complexity and reader capacity, (3) monitoring student perceptions, (4) collaborating to create assessments, (5) using rubrics as assessment tools, (6) providing timely and effective feedback, and (7) analyzing and applying data. In this chapter, we break with the format we established in previous chapters to focus specifically on each of these considerations.

Understanding the Role of Literacy-Based Assessment in the Mathematics Classroom

Assessment is a natural component of the classroom and always has been the primary source of identifying a student's understanding of content. Assessments drive so many elements of learning within our schools and beyond, including determining final grades and influencing college admissions, which often leads schools, media, and society to focus heavily on this educational tool. The importance that assessments bear means that, as an educator, your school leadership, as well as community stakeholders expect you to be up to date on the most recent assessment trends and subscribe to some sort of methodology. This task can be daunting; whether it be assessments *for* learning (Stiggins, 2005) or standards-based grading (Townsley & Wear, 2020), there is always a new idea out there.

While there are many assessment philosophies, the consistent element among most of them is using assessment data to inform instruction (Bailey & Jakicic, 2012). If teams are not using assessments to better identify the next steps for instruction, they are misusing the data those assessments produce. After reviewing students' assessments, teachers may respond with a follow-up lesson by reteaching a concept the next day or by conducting a long-term response involving curriculum revision. The next steps depend on the purpose and format of the assessment. Typically, most assessments fit into one of three categories.

1. **Formative:** These are your in-class assessments that teachers craft to inform themselves and their students of each student's progress toward mastery. These tests may carry score weight but are not the last opportunity for a student to demonstrate their knowledge of a target, standard, and task. Note that depending on your school's testing philosophy, formal progress monitoring (between benchmark periods) may also fall within this category.

2. **Summative:** These assessments are less frequent and serve as an endcap to teaching and practicing a concept. You may administer summative assessments as part of an end-of-unit assessment or even as part of a larger context, like course final exams. That said, as with all team-issued assessments, we encourage you to use results, including summative assessments, to drive continuous improvement in teaching and learning.
3. **Benchmark:** Teams offer benchmark assessments at specific intervals throughout the school year to capture student academic growth from a holistic perspective. Many teams use benchmark assessments to gauge students' knowledge at the start of a semester, school year, or high school career, and then again at the end. These data are typically for tracking students' learning progress and not for determining grades for a particular course. These assessments are often group-administered and nationally normed, meaning that benchmark assessments deliver a percentile comparing each student to others in the same grade or of the same age within a statewide or national pool of students representing all student populations.

Which types of assessments are already a staple in your curriculum? Is your usage of these balanced and appropriate or is one category heavily outweighing your assessment distribution?

As your team approaches the work of literacy assessment within a mathematics context, keep the four critical questions of a PLC in mind. These questions will guide your team discussions and point to your next steps. You may employ these questions holistically to an assessment or to each assessment component. As you do so, trends will emerge, once again pointing you toward the next step, which may range from reteaching a specific lesson in the classroom to reteaching with a small group of students or to engaging one-on-one with a specific struggling learner. Additionally, your team's findings may point toward the next level of inquiry. This may occur in the form of conducting a deeper-dive assessment, aiming your instruction, implementing additional student practice, or conducting further assessment around a specific and more narrow skill that a previous assessment identified as a common relative weakness for a cohort of students.

Matching Text Complexity and Reader Capacity

When crafting literacy-based assessments for the classroom, it is vital that your team design with text complexity and reader capacity in mind. First, team members must know the readers in their classroom and their skill sets. Typically, teams design or choose these assessments to be administered to a large group of students at one time, often electronically, which enables scores to be available immediately. Given RTI (Buffum et al., 2018) and MTSS (National Center on Intensive Intervention, n.d.) have become common practice across the United States, many schools are already conducting this type of benchmarking assessment; however, time and again, we have noted that the results of such benchmarking often go unused. Reach out to your school's reading specialist to learn more about what type of literacy-based benchmarking assessment might already be in place. If you don't have a building-based reading specialist, check with your principal, who should be able to point you in the right direction.

To ensure schools use these results effectively, leadership teams can consider the following questions.

- Does your school conduct benchmark literacy assessments?
- When does benchmark testing occur?
- Which student populations participate?
- Where is the data housed?
- What happens with the data?
- How can this data be shared with core departments?

Note that anyone can start this conversation. Bringing these questions to your leadership is simply one way to initiate and request the support that your team needs and prompt your leadership to partake in reflection.

Because all content-area teachers are teachers of reading, it is imperative that your team accesses students' literacy data. The data will help you scaffold instruction and select materials and texts. Additionally, it will help you get a better understanding of *why* students are struggling or excelling. Understanding the why can clarify if the student needs additional assistance with the reading of the mathematics or support in the calculations or process of mathematics. This will help the instructor know where to intervene.

The goal is to understand each student's *independent*, *instructional*, and *frustration* reading levels, as follows.

- **Independent:** This is the level of text that a student can navigate independently with fluency and solid comprehension. The student does not need any supports or scaffolds when engaging with this text. A student still learns and gains new insights with these texts, but more likely through learning new background knowledge and making inferences, not necessarily by expanding their working vocabulary. Matching a student with an independent-level text is a great option for independent-reading tasks.
- **Instructional:** This level of text is an appropriate level for reading materials that teams use within the classroom to challenge students. This level fosters literacy skill growth when the teacher provides support during the comprehension task. A student's instructional reading range falls within the independent and frustration levels and can often span multiple grades of text complexity; for example, a student's instructional reading range may span grades 8–10.
- **Frustration:** A text at a student's level of frustration is one where the student will have large gaps in understanding due to not having sufficient background knowledge, vocabulary skills, or both to decipher the meaning of the text. Employing such a text in the classroom is best reserved for one-on-one or small-group teacher-guided reading with heavy vocabulary and background knowledge frontloading (see strategies in chapter 4, page 63).

The purpose of knowing this information is to use it as you select texts and build your curriculum and craft your assessments. There should be a match in text complexity between lessons and assessments. If your team knows that you have students who have a sixth-grade instructional reading level, then the students' frustration will be evident when you make them use a textbook with a tenth-grade reading level. We see this kind of mismatch between student reading levels and textbook complexity quite often when teachers are unaware of the drastic mismatch that often happens with readers and unsupported text. The result is a missed opportunity for skill and vocabulary development, leading to frustrated students and closed books.

As teachers, we have honed and developed our own strong literacy skills, and as we are teaching high school courses, we often assume that our students are on the same track. However, the reality is that many of our students cannot work with grade-level materials yet. This reality is sometimes masked by content confusion, when the root of the issue is in fact weakened literacy skills crippled by poor vocabulary and limited background knowledge. By sticking only with grade-level materials, your team is missing opportunities to build the vocabulary and background knowledge that your students need to move toward grade-level reading comprehension.

Although many of your textbooks might be well above the instructional reading levels of your students, the solution cannot be to abandon a mandated textbook and summarize it for your students. We have seen this trend in the past as teachers moved away from books and toward slide-based presentations (such as PowerPoint or Google Slides) as a means of text summary. Instead, we encourage you to supplement with engaging, relevant texts that will help build background knowledge and build student vocabulary and by frontloading challenging text with prereading strategies as scaffolds. In doing so, your students are more prepared to tackle that challenging textbook in class with you to support them—not at home left with their own frustrations.

Consider a particular unit you teach. Do you see students struggle with text at a specific point? What can your team do to scaffold the complex problems and build background knowledge around the problems? What vocabulary do the students need to specifically understand with instructional-level materials?

Monitoring Student Perceptions

Assessing students' perceptions of their literacy skills and practices provides insight into how students approach text and navigate comprehension pitfalls. In 2018, our school's literacy team decided that as part of our school improvement work, it wanted to survey students to see how they engage in text for academic and nonacademic purposes. We administered a survey in student-friendly language that asked about specific strategies and the scenarios in which they would apply these skills. On a whole-school global level, the data were certainly interesting, but on a department level, the results had more meaning. For example, annotation is something that the English and literacy departments viewed as essential to mastering their content, although few students reported that they do this as a during-reading activity. Additionally, most students reported that when they feel confused while reading a text, they reread. This sounds great, but it made us wonder, do students know how to reread? Or are they simply just taking the same faulty approach that led to their confusion in the first place? These two examples have led us to the important next steps: we need to model these fix-up strategies for students and scaffold these self-remediation skills.

Collaborating to Create Assessments

This work of designing assessments is best done as part of a collaborative team. Doing so, in some capacity, helps teams write common assessments that accurately measure student growth with content (learning goals) and process mastery while also producing useful data for team discussion. It is important to assess student growth in both these domains. It's obviously important for students to achieve mastery of learning goals, but process mastery helps students achieve deeper understandings and move toward proficiency with disciplinary literacy.

The texts teams select to use for assessments should speak directly to your team's curriculum standards and echo its formative text tasks. In crafting reading skill inventories, your team must ensure that the text topic mirrors or extends the current course content. So if you are studying a unit on slope intercept form and you are using multiple strategies to teach this concept, then all the strategies should be carried over to the next concept of graphing slope. We want to continue the strategies and reinforce them as we move forward with student learning.

The text and tasks of the reading skill assessment, both formative and summative, should mirror those that you will work on throughout your curriculum. Additionally, the questions and analysis tasks should directly speak to the literacy outcomes that your team's discipline requires. Make sure that the texts are also within an appropriate instructional range for the students your team is testing.

Typically, teams should collaboratively design two or three assessments that target the same skills and either give them as a preassessment and postassessment to monitor prior knowledge and then student mastery of learning goals or as a preassessment, mid-unit assessment, and postassessment to additionally check in on student progress. It is important for teachers to give the preassessment prior to introducing any skills to establish baseline data and help determine the level of support students need. Teachers should administer the next assessment after there has been time for thorough instruction, guided practice paired with checks for understanding, and independent application. These assessment results will provide ample discussion opportunities for your team to shape how each team member teaches power standards (see chapter 1, page 19).

Know that an assessment does not need to take on one specific format. In fact, using a variety of assessment methods conveys better the extent to which students can transfer and transport content-area literacy skills from one context to the next. It is also essential that your assessments are balanced, meaning they should not all be formative or summative. Additionally, depending on the structure of your course, not all assessments need to be common. Teams may choose to assess common core content targets, but individual teachers can determine a cluster of different skills to assess depending on the proficiency students show in each classroom. This will foster a truly differentiated learning environment that aligns with the loose-tight nature of a PLC (DuFour et al., 2024). Further, although formative assessments should be your primary tool, this does not necessarily mean that all formative assessments should receive a grade. What matters most is that you are continually providing students with opportunities to demonstrate their learning in incremental phases, leading toward summative assessment tasks. In turn, your team collects data it can use to drive instruction, including intervention and extension.

Strong, literacy-based assessments prompt students to employ the same literacy skills they have been practicing in prereading, during-reading, and postreading activities but with a new text that is familiar in format. If you are teaching struggling readers, you may even prompt students to employ the same scaffolds you taught and modeled in class; over time, you can and should eliminate these scaffolds as students gain independence and autonomy as disciplinary readers. In such a scenario, the strategies we outline in this book not only serve as your strategy scaffold for teaching a specific literacy skill but also as your assessment of the student's ability to apply that learned skill to new mathematical contexts.

Using Rubrics as Assessment Tools

Rubrics have long been a useful tool for teachers to provide students with specific, targeted feedback on their work. When given to students for self-reflection, rubrics can serve as a powerful self-assessment tool. Students are then able to evaluate their work steps and processes to see if they are correct. There are many different rubric variations based on the specific target or task the rubric focuses students on. We've seen rubrics commonly used to detail a *minimum* of twenty descriptor boxes, each indicating a level of correctness. Although we sometimes still find these useful for a final, summative assessment, our team has moved toward using single-target rubrics as part of the formative and reflective learning processes (figure 7.1).

Task
The Brooktown soccer team is selling boxes of cookies.

Chocolate chip cookies are \$5 a box.	Peanut butter cookies are \$4 a box.	Sugar cookies are \$3 a box.

Answer the following questions and show your work.

1. Jasmine sold 2 boxes of sugar cookies. How much money did she raise?
2. Sean sold 4 boxes of peanut butter cookies and 4 boxes of chocolate chip cookies. How much money did he raise?
3. Mr. Jones spent \$25 on chocolate chip cookies. How many boxes did he buy?
4. Mica's mother has \$18. How many boxes of sugar cookies can she buy?

Rubric		
This task requires students to work with multiplication and division in a real context. Credit for aspects of performance are assigned as follows:	**Points**	**Section Points**
1. Gives correct answer: \$6.00 and shows some correct work, such as 3 + 3 = 6	1	1
2. Gives correct answer: \$36.00	1	2
Shows work, such as 4 × 4 and 4 × 5	1	
3. Gives correct answer: 5	1	2
Shows work, such as 25 ÷ 5	1	
4. Gives correct answer: 6	1	2
Gives a correct explanation such as, The most her mother can buy is 6 boxes because I counted by 3s and she can buy 6 but doesn't have enough for 7 or, I counted 3, 6, 9, 12, 15, 18, 21. She doesn't have 21.	1	
Total Points		**7**

Figure 7.1: Rubric to provide timely and effective feedback.

This type of rubric will work well with just about any standard that involves a process, making it a good fit for assessing academic standards. Creating a single-standard rubric allows for in-depth reflection and progress analysis of critical skills. This can also be done by creating the rubric together as a class. In this case, the teacher could provide the learning target criteria. The various levels of working toward and exceeding the standard could be charted during the class discussion. Taking the time to craft the rubric together ensures that students understand the objective and will better be able to navigate and monitor their progress toward it.

Providing Timely and Effective Feedback

Feedback is essential for student growth. We're sure everyone remembers a time when they were a student and worked hard on an assignment or assessment, such as a research paper or a particularly critical final exam. As soon as you handed in your work, all you could think about that day was, How did I do? The very next day, you probably walked into class, just hoping that you would have an answer, any answer, to that question so that your nerves could relax.

We are not suggesting that your teams need to provide immediate feedback to students on every assignment or assessment. What matters is finding a balance in the feedback process so that team members prompt student reflection and growth while not getting bogged down in the infinite world of feedback. When teachers are unbalanced in their approach, they often find themselves in a situation where they spend more time providing feedback than the student did completing the assignment. This is not a productive feedback system! Our experience finds that most secondary-level students are simply not able to recognize their own errors, and if they can, they likely don't have the tools to correct their work without some form of prompting. This prompting can happen via a number of discourses, both formal and informal; however, it is critical to recognize that the longer students go without feedback, the less likely they are to reflect on that feedback and learn from it.

While some assignments, like a major research paper, certainly require some time for team members to provide feedback, there are many practical ways to deliver quick and timely feedback to students. As you saw in figure 7.1 (page 115), a well-crafted feedback tool, like a rubric, will save you quite a bit of time while also providing students with specific points to reflect on and grow. Rubrics make the end goal obvious and measurable in a concrete way while quickening the process and making feedback discrete instead of ambiguous. Having the specific success criteria rubrics breaks down the larger learning targets, making it possible for learners to identify where to focus their efforts to improve overall. Likewise, when introducing a new concept, checking for understanding with a quick exit slip or entrance slip the next day and providing feedback with a simple + or – is also an efficient way of letting students know if they are on the right track.

There are limitless possibilities for effective feedback, so be creative in your feedback methods. Did you ever have a teacher who asked you and your student peers to trade papers as a means to quickly score an in-class quiz? Try this alternative: have students keep their quizzes and put all utensils away except for a pen that is a different color than the one they used to take the test. Then, go over the correct answers, including the reasoning and evidence necessary to support why an answer is correct while having students adjust their own errors. Not only have you provided timely feedback, but you were able to help students reflect and identify misconceptions and have also avoided the shame that some students feel when exposing their work to others. The benefit of timely feedback for students is clear—less time being confused and making the same errors over and over again. As this kind of feedback makes any student error trends clear, follow this data gathering with a minilesson to refocus student learning and address those misconceptions. Even for those who demonstrate understanding on the first try, the minilesson will help to reinforce that understanding. Finally, as we cover next, collecting feedback gives you data to return to your team regarding any learning trends that you identified.

Performing Item Analysis

After an assessment, it is critical to not only provide students with timely feedback but also bring samples of student assessments back to your collaborative team for further discussion. In terms of analysis, your team should design assessments to gather data that are aligned with the team's learning targets and process standards. When doing this work during the assessment-design phase, there are two major lenses (perspectives) to employ while taking a deep dive into these data: (1) the whole class and (2) a sample pack—collections of a few assessments from different student populations. For example, your team might collect assessments based on students' reading levels, for students receiving specific forms of intervention, for students who are learning English, and so on.

When assessment structures align with a team's learning targets, gathering data for a whole class (or even multiple classes) is very easy to do because they can be put together and aggregated. There are a number of tools that can support this process, such as Mastery Manager (www.masterymanager.com) and Google Classroom (https://classroom.google.com), and provide your team with a useful way to look for overarching trends of correct and incorrect responses.

The sample-pack method works best when your team wants to go deeper than just looking at whole-class trends but rather seeks to zoom in on the progress of more specific student groups. Taking the time to thoroughly explore a subset of assessment results takes longer; however, it's an ideal approach for identifying where students may have made a wrong turn in the process, logic, or application.

When working with our team, we often employ both whole-class and sample-pack lenses. First, we look at final answer scores for an entire class or population, then also analyze a sample pack to further identify where students made specific mistakes. Consequently, these combined data guide our next steps of reteaching. Figure 7.2 provides guiding questions for teams to conduct data analysis for both a whole class and via a sample pack.

1. Whole Class
    - As a whole, how did students perform? Are there any tasks or questions that proved problematic for the class as a whole? If yes, review how these topics or tasks were covered in your curriculum and teaching.
    - Do you see any general trends among student assessment data for multiple sections (classes) of the same course?
    - To what can you attribute student successes and challenges?
2. Sample Pack
    - What variations do you see in terms of performance by the group?
    - Is there a need for additional scaffolds for specific students demonstrating a common need?

Figure 7.2: Data-analysis perspectives.

Visit ***go.SolutionTree.com/literacy*** *for a free reproducible version of this figure.*

Once your team has collected and analyzed data, the next critical step is to decide what to do with what the data communicate. Is the student struggling with mathematics literacy or numeracy? Different data points and data sets have different purposes. It may seem clear what to do after conducting an analysis of formative data, such as identifying where students need more instruction and using that knowledge to revisit a correlated learning target with a more scaffolded process. Benchmark or normative data, on the other hand, sometimes leave teams wondering what to do with them. In our collaborative team, we always start our discussion with the obvious: the fact that these data likely confirm or add evidence and clarification to what we have already observed of student proficiency. If this isn't the case, then we ask, What were the surprises? This task can be daunting, especially if the data confirm that you may have readers in your class operating at a four-grade (or more) deficit from the content you are teaching. Don't let these data points thwart you; again, they only confirm what you probably already suspected. Even if they come as a surprise, the data are critical information for you and your team to have. As your team moves forward, these data give it the necessary marching orders for how to support students in developing a wide array of skills.

Teachers will not be able to craft a differentiated lesson for each student in the room, nor employ multiple texts and primary sources on the same topic each day. However, they can and should look for opportunities to do so throughout a unit, and mathematics provides a great platform for this kind of work. We selected and crafted the strategies in chapters 3–5 to help you do just this.

As your team deploys these strategies, consider how creating student learning groups based on assessment data can be a productive use of the data. Many online assessment systems, such as the assessment tools at Renaissance Learning (www.renaissance.com), have tools integrated as part of their reporting features that allow teams to easily group students and will even generate skill data for groups. Whether using an online data tool or old-school paper and pen, here are some basic grouping methods. Which one team members choose ultimately depends on the task and purpose.

- **Homogeneous groups:** Group students based on similar assessment results to focus on developing a group-specific skill or to master a group-specific content goal (students with like scores).
- **Linear groups:** Create a list of students from highest to lowest score and divide them into groups accordingly.
- **Heterogeneous groups:** Group students with highly variable skill sets or proficiency levels into one group so that they can benefit from each other's areas of strength. The key to this grouping is to ensure students work as a community for mutual benefit, rather than (as we sometimes see) one or two students working while the others are either lost or copying the work without understanding it. Another way to distribute groups heterogeneously is to list the students in a vertical row from highest score to lowest score. Then, cut the list in half, and place the two lists side-by-side. Each pair of names then works as a paired group.

While there are many more approaches your team could take, what's important is to think creatively about how you can group students in a way that allows them to grow and benefit from collaboration and collective strengths. While randomly grouping students by counting off in class or organizing based on birthdays may provide variety, these methods don't fit any specific purpose or further skill development and content mastery.

Wrapping Up

Meaningful assessment requires a great deal of intention and reflection from your team. While it is the student's responsibility to apply the skills they have developed, a team's role is to create assessments that intentionally and accurately mirror the curriculum and skills each team member teaches. These kinds of intentional assessments and reflection lead to responsive teaching for the entire PLC. This act of inquiry fosters the thoughtful selection and development of tools, texts, and tasks that help students progress toward being critically literate and possessing the disciplinary literacy skills necessary to not only grow within your content classroom but also in the post–high school world.

Collaborative Considerations *for* Teams

- Does your team have a robust assessment cycle that balances formative and summative assessments?
- Does your team take student reading ability into account when selecting texts for use in assessments? Does the team have an awareness of text and task complexity?
- Has your team collaborated on how to best share feedback with students that will impact future learning and growth?

EPILOGUE

Building capacity for collaboration among mathematics teachers and literacy experts is truly the strongest catalyst for supporting student growth across all disciplines, and often, it is where this work begins. In truth, schools often look to the English teachers to take the first step toward this kind of collaboration; it is a unique opportunity. However, when literacy experts and mathematic experts get into one room and collaborate it is extremely powerful. As schools commit to exemplary instruction to support the growth of every learner in every discipline, this is something that educators with the power of a PLC behind them must not overlook. Regardless of the structure of teams in your PLC, whether teams are organized based on discipline, vertically, or across departments, this work always begins with teaming teachers who are focused on improving their own capacity to impact student learning and scaffold critical literacy skills—those that transfer content and task.

While this book speaks directly to mathematics experts who want to hone their teaching of literacy in their classroom, this is one installment in a series that will support teacher collaboration and strategic literacy-infused teaching in all content areas. Each text in this series focuses on building a common language and literacy thought partners across all disciplines. Ultimately, the literacy skills developed in grades 6–12 students throughout their academic career will greatly impact their ability to think critically and their overall readiness for college and career.

APPENDIX:

REPRODUCIBLES

Reading Comprehension Process

Did I . . . ?	Strategic Comprehension Step	Before, During, or After Reading
☐	**Preview text**, ask questions, and make predictions.	**Before:** Focus and get ready to read.
☐	**Recall** what you already know about the topic.	
☐	Set a **purpose** for reading.	
☐	Make a **notetaking plan** for remembering what's important.	
☐	Define **key concepts** and important vocabulary whenever possible.	
☐	Keep your **purpose** for reading in mind.	**During:** Stay mentally active.
☐	**Make meaning** by: • Asking questions • Putting the main ideas into your own words • Visualizing what you read • Making notes to remember what's important • Making connections between the text and people, places, things, or ideas	
☐	**Be aware** of what's happening in your mind as you read. Consider: • Am I focused or distracted? • Do I need to go back to a part I didn't get and reread it? • What are my reactions to what I am reading?	
☐	**Reflect on** what you've read. Consider: • Did I find out what I needed or wanted to know? • Can I summarize the main ideas and important details in my own words? • Can I apply what I have learned? • Can I talk about or write about what I have learned?	**After:** Check for understanding.

Content-Standard Analysis Tool

Content Standard	Artifacts		Power Standard
	Lesson Plans and Course Materials	Student-Generated Evidence of Standard	☐
Unpacked (process standards):			Opportunities and Next Steps
	Strengths and Needs	Strengths and Needs	

Content Standard-Analysis Steps (Use this with your team to guide collaborative discussions.)

1. **Unpack:** What is the standard really saying? Put the standard in your own words and determine the individual process standards inherent within the content standard.
2. **Identify artifacts:**
    - Lesson plans and course materials—
        - How is the standard being taught?
    - Student-generated evidence—
        - How are students demonstrating mastery?
3. **Assess the strengths and weaknesses of these artifacts:** Where is there room for improvement?
4. **Identify power standards:** Which content outcomes are the most essential to your content area?
5. **Identify next steps:** As your team's work continues, identify potential teaching and learning opportunities.

Review Tool to Discern Whether Text Tasks Match Complex Disciplinary Literacy Demands

1. Does the content match your defined content power standards?
   If not, what is missing?
2. Are these texts at an appropriate text-complexity level for your readers?
3. Are your text tools varied to address the different types of content or texts associated with your discipline?
4. Once you have gathered your collection of materials, what scaffolds will your students need to comprehend the content successfully?
5. What scaffolds will your students need in order to transfer knowledge to new contexts and applications?

Fix-Up Strategies Bookmark

FIX-UP STRATEGY BOOKMARK

★ Make a connection.

★ Make a prediction.

★ Stop and think about what you are reading.

★ Ask yourself a question.

★ Reflect in writing on what you've read.

★ Visualize: Stop and create mental images of the concepts described in the text.

★ Use print conventions: Examine directions and emphasized words to identify important ideas in the text.

★ Retell what you've read.

★ Reread: If something doesn't make sense, reread for clarity.

★ Notice patterns in text.

★ Slow down or speed up: Adjust your reading to slow down or speed up depending on your level of understanding.

Mathematics Process Log

Mathematics Process Log: **Question:** **Extra challenge:**	
Presolving:	
Presolving:	
Presolving:	
Presolving:	
Solving:	**Figure out the data:**
Solving:	**Put the numbers in the equation:**
Solving:	**Solve the problem:**
Solving:	**Write an answer:**
Postsolving:	**Check your answer:**

Source: Adapted from Buehl (2017).

Text-Dependent Questioning Strategy and Graphic Organizer

Task to solve:		
Text-Dependent Questioning	**Questioning the Problem: General**	**My Focused Questions**
Said what?	• What is the problem saying? • What is the problem telling me? • What does the problem say I need to clarify? What can I do to clarify what the problem asks? What does the problem assume I already know? • What is the problem asking me to do?	
Did what? What information does the problem provide?	• What types of information are given [number, statistic, description, relationship, or visual]? • How does vocabulary give clues to solve the problem? • How does the text structure show mathematical patterns, connections, or relationships? • How does the problem indicate what it is asking to be solved? • How can I identify and efficiently and appropriately solve the problem? • What concepts, formulas, and ideas does the problem explain or reveal?	
So What? What connections can you make?	• Can I solve the problem? • Can I verify that my solution is accurate? • What does the problem want me to understand? • What connection is the problem revealing? • What is the problem arguing, and what support (evidence and reasoning) does the problem present? • How can I represent the solution (graphically, algebraically, numerically)? • Does my answer make sense?	
Now what? Now, what can you do with your understanding of the problem?	• How does the problem connect with your previous knowledge or experience? • How does the problem influence or change your thinking? • What implications can you draw from the solutions to the problem? • Do more efficient options exist for solving the problem? • How does this relate to the learning targets for the course?	

Source: Eric Anderson, 2017. Used with permission.

Frayer Model Graphic Organizer

Directions: Write down a vocabulary word or concept that was critical to your reading and explain it using the included prompts.

Definition (what it is):	**Characteristics:**
Vocabulary Word:	
Examples or pictures:	**What it is not:**

Predicting and Confirming Activity

Problem:

Prediction	Confirmed (+)	Not Confirmed (–)	Support

Data-Analysis Perspectives

1. Whole Class

    - As a whole, how did students perform? Are there any tasks or questions that proved problematic for the class as a whole? If yes, review how these topics or tasks were covered in your curriculum and teaching.
    - Do you see any general trends among student assessment data for multiple sections (classes) of the same course?
    - To what can you contribute student successes and challenges?

2. Sample Pack

    - What variations do you see in terms of performance by the group?
    - Is there a need for additional scaffolds for specific students demonstrating a common need?

REFERENCES AND RESOURCES

Abosalem, Y. (2016). Assessment techniques and students' higher-order thinking skills. *International Journal of Secondary Education, 4*(1), 1–11. Accessed at https://pdfs.semanticscholar.org/81e4/0f2f1321180b6acf5de0d53b7f05251ba030.pdf on August 12, 2019.

Anderson, L. W., & Krathwohl, D. (Eds.). (2001). *A taxonomy for learning, teaching, and assessing: A revision of Bloom's taxonomy of educational objectives*. Boston: Allyn & Bacon.

Aronson, E., & Patnoe, S. (1997). *The jigsaw classroom: Building cooperation in the classroom* (2nd ed.). New York: Addison Wesley Longman.

Bailey, K., & Jakicic, C. (2012). *Common formative assessment: A toolkit for Professional Learning Communities at Work*. Bloomington, IN: Solution Tree Press.

Bloom, B. S. (Ed.). (1956). *Taxonomy of educational objectives: The classification of educational goals; Handbook I: Cognitive domain*. New York: McKay.

Buehl, D. (2017). *Developing readers in the academic disciplines* (2nd ed.). Portland, ME: Stenhouse.

Buffum, A., Mattos, M., & Malone, J. (2018). *Taking action: A handbook for RTI at Work*™. Bloomington, IN: Solution Tree Press.

Buffum, A., Mattos, M., & Weber, C. (2012). *Simplifying response to intervention: Four essential guiding principles*. Bloomington, IN: Solution Tree Press.

Columbia University Teachers College. (2005, June 8). *The academic achievement gap: Facts & figures*. Accessed at www.tc.columbia.edu/articles/2005/june/the-academicachievement-gap-facts--figures on July 15, 2024.

Conzemius, A. E., & O'Neill, J. (2014). *The handbook for SMART school teams: Revitalizing best practices for collaboration* (2nd ed.). Bloomington, IN: Solution Tree Press.

Coons, C. (2022). *What is a paragraph proof?* Accessed at https://study.com/academy/lesson/what-is-a-paragraph-proof-definition-examples.html on May 23, 2023.

Csikszentmihalyi, M. (2009). *Flow: The psychology of optimal experience*. New York: HarperCollins.

Duke, N. K., & Pearson, D. (2002). Effective practices for developing reading comprehension. In A. E. Farstrup, S. J. Samuels, & J. Samuels (Eds.), *What research has to say about reading instruction* (3rd ed., pp. 205–242). Accessed at www.researchgate.net/publication/303174858_What_research_has_to_say_about_reading_instruction.

DuFour, R. (2004). What is a "professional learning community?" *Educational Leadership, 61*(8), 6–11. Accessed at www.siprep.org/uploaded/ProfessionalDevelopment/Readings/PLC.pdf on July 15, 2024.

DuFour, R., DuFour, R., Eaker, R., Many, T. W., Mattos, M., & Muhammad, A. (2024). *Learning by doing: A handbook for Professional Learning Communities at Work* (4th ed.). Bloomington, IN: Solution Tree Press.

Fang, Z., & Coatoam, S. (2013). Disciplinary literacy: What you want to know about it. *Journal of Adolescent and Adult Literacy, 56*(8), 627–632. Accessed at http://www.jstor.org/stable/41827916 on May 23, 2023.

Ferlazzo, L., & Sypnieski, K. H. (2018). *Activating prior knowledge with English language learners*. Accessed at www.edutopia.org/article/activating-prior-knowledge-english-language-learners on August 26, 2020.

Ferriter, W. M. (2020). *The big book of tools for collaborative teams in a PLC at Work®*. Bloomington, IN: Solution Tree Press.

Frayer, D. A., Frederick, W. C., & Klausmeier, H. J. (1969). *A schema for testing the level of cognitive mastery*. Madison, WI: Wisconsin Center for Education Research.

Gabriel, R., & Wenz, C. (2017). Three directions for disciplinary literacy. *Educational Leadership, 74*(5), 8–14. Accessed at https://researchgate.net/publication/316926851_Three_Directions_for_Disciplinary_Literacy on May 23, 2023.

Gambrell, L. B. (2011). Seven rules of engagement: What's most important to know about motivation to read. *The Reading Teacher, 65*(3), 172–178.

Garmston, R., & Wellman, B. (1998). Teacher talk that makes a difference. *Educational Leadership, 55*(7), 30–34.

Goodwin, B. (2017). Research matters: Helping students develop schemas. *Educational Leadership, 75*(2), 81–82.

Hussar, B., Zhang, J., Hein, S., Wang, K., Roberts, A., Cui, J., et al. (2020, May). *The condition of education 2020* (NCES 2020,144). U.S. Department of Education. Washington, DC: National Center for Education Statistics. Accessed at https://nces.ed.gov/pubs2020/2020144.pdf on May 23, 2023.

Kise, J. A. G. (2021). *Doable differentiation: 12 strategies to meet the needs of all learners*. Bloomington, IN: Solution Tree Press.

National Center on Intensive Intervention. (n.d.). *Intensive intervention & MTSS*. Accessed at https://intensiveintervention.org/special-topics/mtss on May 23, 2023.

National Governors Association Center for Best Practices & Council of Chief State School Officers (NGA & CCSSO). (2010). *Common Core State Standards for Mathematics*. Accessed at https://learning.ccsso.org/wp-content/uploads/2022/11/ADA-Compliant-Math-Standards.pdf on May 23, 2023.

ProLiteracy. (n.d.). *U.S. adult literacy facts* [Brochure]. Accessed at https://proliteracy.org/Portals/0/pdf/PL_AdultLitFacts_US_flyer.pdf?ver=2016-05-06-145137-067 on May 23, 2023.

Rebell, M. A. (2008). Equal opportunity and the courts. *Phi Delta Kappan, 89*(6), 432–439.

RTI Action Network. (n.d.). *Tiered instruction/intervention*. Accessed at http://rtinetwork.org/essential/tieredinstruction on July 15, 2024.

Rumelhart, D. E. (1980). Schemata: The building blocks of cognition. In R. J. Spiro, B. C. Bruce, & W. F. Brewer (Eds.), *Theoretical issues in reading comprehension: Perspective and cognitive psychology, linguistics, artificial intelligence, and education* (pp. 33–58). Hillsdale, NJ: Erlbaum.

Shanahan, T., & Shanahan, C. (2008). Teaching disciplinary literacy to adolescents: Rethinking content-area literacy. *Harvard Educational Review, 78*(1), 40–59, 279. Accessed at https://dpi.wi.gov/sites/default/files/imce/cal/pdf/teaching-dl.pdf on July 15, 2024.

Siegel, M., & Fonzi, J. M. (1995). The practice of reading in an inquiry-oriented mathematics class. *Reading Research Quarterly, 30*(4), 632–673.

Stephens, D., Morgan, D. N., DeFord, D. E., Donnelly, A., Hamel, E., Keith, K. J., et al. (2011). The impact of literacy coaches on teachers' beliefs and practices. *Journal of Literacy Research, 43*(3), 215–249.

Stiggins, R. (2005). From formative assessment to assessment FOR learning: A path to success in standards-based schools. *Phi Delta Kappan, 87*(4), 324–328.

Tecca, L., Weaver, J., & Chen, J. (2023). The importance of implementing literacy strategies in a mathematics classroom. *Honors Projects*. 904. Accessed at https://scholarworks.bgsu.edu/honorsprojects/904 on May 5, 2024.

ThinkCERCA. (n.d.a). *Our story*. Accessed at https://thinkcerca.com/story on May 23, 2023.

ThinkCERCA. (n.d.b). *Why argumentation? Our research-based approach*. Accessed at https://thinkcerca.com/argumentation-research-based-approach on January 18, 2019.

Tovani, C. (2000). *I read it, but I don't get it: Comprehension strategies for adolescent readers*. Portland, ME: Stenhouse.

Townsley, M., & Wear, N. L. (2020). *Making grades matter: Standards-based grading in a Secondary PLC at Work*. Bloomington, IN: Solution Tree Press.

Urquhart, V., & Frazee, D. (2012). *Teaching reading in the content areas: If not me, then who?* (3rd ed.). Alexandria, VA: Association of Supervision and Curriculum Development.

U.S. Department of Education, National Center for Education Statistics, National Assessment of Educational Progress. (n.d.). *NAEP reading report card*. Accessed at www.nationsreportcard.gov/reading_2017?grade=4 on September 4, 2019.

Weiss, M. K., & Moore-Russo, D. (2012). Thinking like a mathematician. *The Mathematics Teacher, 106*(4), 269–273. https://doi.org/10.5951/mathteacher.106.4.0269

INDEX

T

V

W

Reading and Writing Strategies for the Secondary English Classroom in a PLC at Work®
Edited by Mark Onuscheck and Jeanne Spiller
Close literacy achievement gaps across grades 6–12. Part of the *Every Teacher Is a Literacy Teacher* series, this resource highlights how English language arts teachers can work collaboratively to combat literacy concerns and improve student skill development.
BKF904

Reading and Writing Strategies for the Secondary Science Classroom in a PLC at Work®
Edited by Mark Onuscheck and Jeanne Spiller
Equip your students with the literacy support they need to think like scientists. Written by a team of experienced educators, this book provides practical literacy-based strategies specifically for science teachers of grades 6–12.
BKF907

Reading and Writing Strategies for the Secondary Social Studies Classroom in a PLC at Work®
Edited by Mark Onuscheck and Jeanne Spiller
Prepare middle school and high school students to read, write, and think like social studies experts and historians. Part of the *Every Teacher Is a Literacy Teacher* series, this resource details how teachers can work collaboratively to support literacy development and social studies learning.
BKF908

The New Art and Science of Teaching Reading
Julia A. Simms and Robert J. Marzano
The New Art and Science of Teaching Reading presents a compelling model for reading development structured around five key topic areas. More than 100 reading-focused instructional strategies are laid out in detail to help teachers ensure every student becomes a proficient reader.
BKF811

Visit SolutionTree.com or call 800.733.6786 to order.

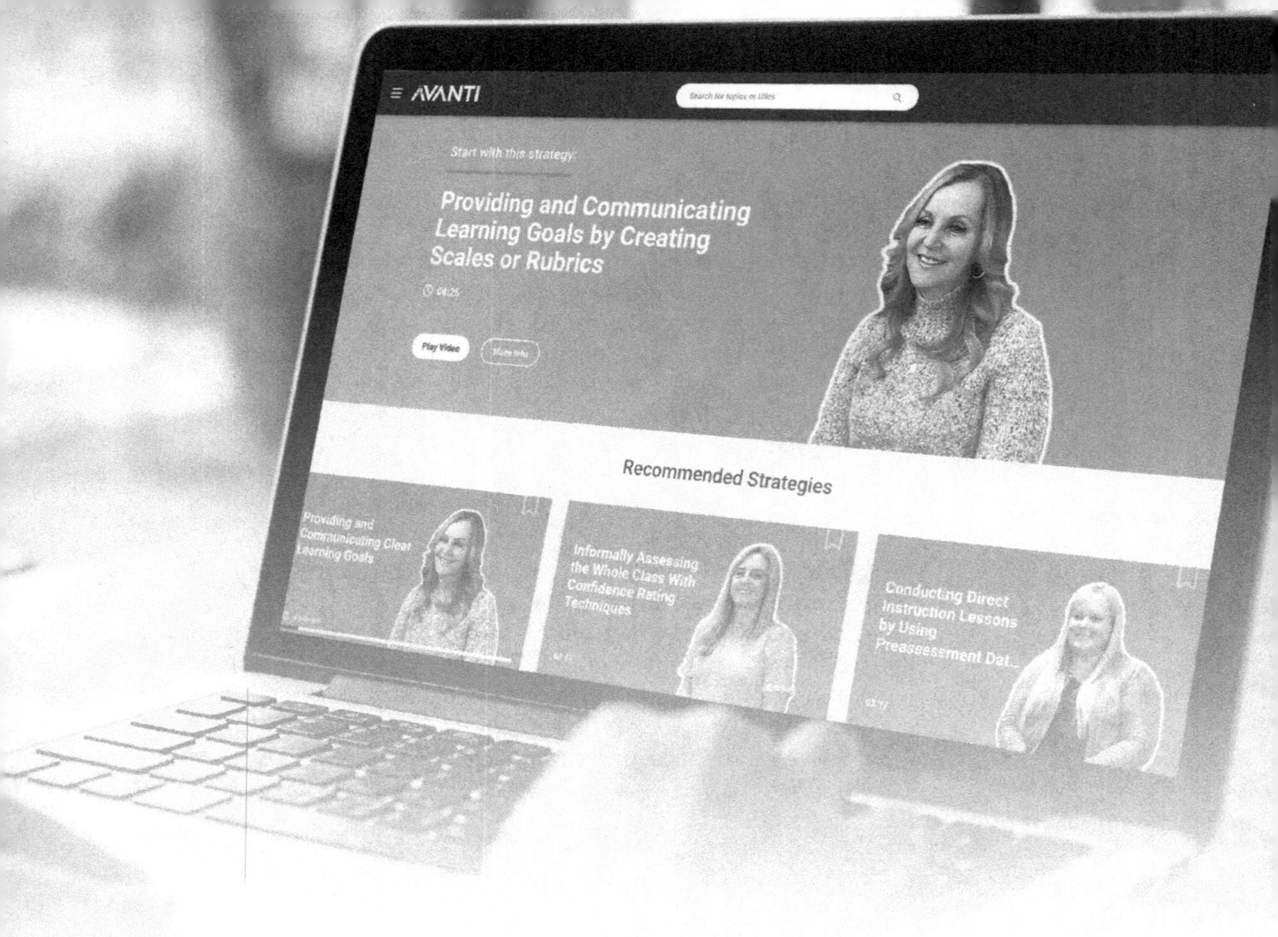